The IMA Volumes in Mathematics and Its Applications

Volume 13

Series Editors
George R. Sell   Hans Weinberger

Institute for Mathematics and Its Applications
IMA

The **Institute for Mathematics and Its Applications** was established by a grant from the National Science Foundation to the University of Minnesota in 1982. The IMA seeks to encourage the development and study of fresh mathematical concepts and questions of concern to the other sciences by bringing together mathematicians and scientists from diverse fields in an atmosphere that will stimulate discussion and collaboration.

The IMA Volumes are intended to involve the broader scientific community in this process.

Hans Weinberger, Director
George R. Sell, Associate Director

IMA Programs

1982–1983 Statistical and Continuum Approaches to Phase Transition

1983–1984 Mathematical Models for the Economics of Decentralized Resource Allocation

1984–1985 Continuum Physics and Partial Differential Equations

1985–1986 Stochastic Differential Equations and Their Applications

1986–1987 Scientific Computation

1987–1988 Applied Combinatorics

1988–1989 Nonlinear Waves

1989–1990 Dynamical Systems and Their Applications

Springer Lecture Notes from the IMA

The Mathematics and Physics of Disordered Media
Editors: Barry Hughes and Barry Ninham
(Lecture Notes in Mathematics, Volume 1035, 1983)

Orienting Polymers
Editor: J. L. Ericksen
(Lecture Notes in Mathematics, Volume 1063, 1984)

New Perspectives in Thermodynamics
Editor: James Serrin
(Springer-Verlag, 1986)

Models of Economic Dynamics
Editor: Hugo Sonnenschein
(Lecture Notes in Economics, Volume 264, 1986)

Martin Schultz
Editor

Numerical Algorithms for Modern Parallel Computer Architectures

With 57 Illustrations

Springer-Verlag
New York Berlin Heidelberg
London Paris Tokyo

Martin Schultz
Research Center for Scientific Computation
Yale University
New Haven, CT 06520
USA

Mathematics Subject Classification (1980): 68C05, 65KXX

Library of Congress Cataloging-in-Publication Data
Numerical algorithms for modern parallel computer architectures / Martin Schultz, editor.
p. cm. -- (The IMA volumes in mathematics and its applications : v. 13)
Papers presented at a workshop organized and sponsored by the Institute for Mathematics and Its Applications of the University of Minnesota.
Bibliography: p.
Includes index.
1. Parallel processing (Electronic computers)--Congresses.
2. Computer architecture--Congresses. 3. Algorithms--Congresses.
I. Schultz, Martin H. II. University of Minnesota. Institute for Mathematics and Its Applications. III. Series.
QA76.5.N79 1988
004'.35--dc19 88-4953

Camera-ready text prepared by the editor.
Printed and bound by R.R. Donnelley & Sons, Harrisonburg, Virginia.
Printed in the United States of America.

9 8 7 6 5 4 3 2 1

ISBN 0-387-96733-8 Springer-Verlag New York Berlin Heidelberg
ISBN 3-540-96733-8 Springer-Verlag Berlin Heidelberg New York

The IMA Volumes in Mathematics and Its Applications

Current Volumes:

Volume 1: Homogenization and Effective Moduli of Materials and Media
Editors: Jerry Ericksen, David Kinderlehrer, Robert Kohn, and J.-L. Lions
Volume 2: Oscillation Theory, Computation, and Methods of Compensated Compactness
Editors: Constantine Dafermos, Jerry Ericksen, David Kinderlehrer, and Marshall Slemrod
Volume 3: Metastability and Incompletely Posed Problems
Editors: Stuart Antman, Jerry Ericksen, David Kinderlehrer, and Ingo Müller
Volume 4: Dynamical Problems in Continuum Physics
Editors: Jerry Bona, Constantine Dafermos, Jerry Ericksen, and David Kinderlehrer
Volume 5: Theory and Application of Liquid Crystals
Editors: Jerry Erickson and David Kinderlehrer
Volume 6: Amorphous Polymers and Non-Newtonian Fluids
Editors: Constantine Dafermos, Jerry Ericksen, and David Kinderlehrer
Volume 7: Random Media
Editor: George Papanicolaou
Volume 8: Percolation Theory and Ergodic Theory of Infinite Particle Systems
Editor: Harry Kesten
Volume 9: Hydrodynamic Behavior and Interacting Particle Systems
Editor: George Papanicolaou
Volume 10: Stochastic Differential Systems, Stochastic Control Theory and Applications
Editors: Wendell Fleming and Pierre-Louis Lions
Volume 11: Numerical Simulation in Oil Recovery
Editor: Mary Fanett Wheeler
Volume 12: Computational Fluid Dynamics and Reacting Gas Flows
Editors: Bjorn Engquist, Mitchell Luskin, and Andrew Majda
Volume 13: Numerical Algorithms for Modern Parallel Computer Architectures
Editor: Martin Schultz
Volume 14: Mathematical Aspects of Scientific Software
Editor: J.R. Rice

Forthcoming Volumes:

1986–1987: *Scientific Computation*

Mathematical Frontiers in Computational Chemical Physics

CONTENTS

Foreword ix
Preface xi

Parallelization of Physical Systems on the BBN Butterfly: Some Examples 1
William Celmaster

Solution of Large Scale Engineering Problems Using a Loosely Coupled Array of Processors 11
E. Clementi, D. Logan, V. Sonnad

A Look at the Evolution of Mathematical Software for Dense Matrix Problems over the Past Fifteen Years 29
J. J. Dongarra and D. C. Sorensen

A Graphical Approach to Load Balancing and Sparse Matrix Vector Multiplication on the Hypercube 37
Geoffrey C. Fox

A Review of Automatic Load Balancing and Decomposition Methods for the Hypercube 63
Geoffrey C. Fox

Adaptive Methods for Hyperbolic Problems on Local Memory Parallel Processors 77
William Gropp

Systolic Algorithms for the Parallel Solution of Dense Symmetric Positive-Definite Toeplitz Systems 85
Ilse C. F. Ipsen

Ensemble Architectures and their Algorithms: An Overview 109
S. Lennart Johnsson

Problem Decomposition and Communication Tradeoffs in a Shared-Memory Multiprocessor 145
Thomas J. LeBlanc

Domain Decomposition Preconditioners for Elliptic Problems in Two and Three Dimensions 163
Joseph E. Pasciak

Methods for Automated Problem Mapping 173
Joel Saltz

Block Algorithms for Parallel Machines 197
Robert Schreiber

The LCAP System: An Example of a Multicomputer 209
Martin H. Schultz

A Block QR Factorization Scheme for Loosely Coupled Systems of Array Processors 217
Charles Van Loan

FOREWORD

This IMA Volume in Mathematics and its Applications

NUMERICAL ALGORITHMS FOR MODERN PARALLEL COMPUTER ARCHITECTURES

is in part the proceedings of a workshop which was an integral part of the 1986-87 IMA program on SCIENTIFIC COMPUTATION. We are grateful to the Scientific Committee: Bjorn Engquist (Chairman), Roland Glowinski, Mitchell Luskin and Andrew Majda for planning and implementing an exciting and stimulating year-long program. We especially thank the Workshop Organizer, Martin Schultz, for arranging a broadly based program in the area of computer architecture.

George R. Sell

Hans Weinberger

PREFACE

It seems likely that parallel computers will have major impact on scientific computing. In fact, their potential for providing orders of magnitude more memory and cpu cycles at low cost is so large as to portend the possibility of completely revolutionizing our very concept of the outer limits of scientific computation. It is now conceivable that computational modeling and simulation will play a major role (perhaps even THE major role) in future advances in the technology of engineering and scientific research and development. Along with traditional theory (mathematics) and experiments it will form a triad of approaches to these technical problems.

The realization of this potential will require major advances in multiprocessor architectures, parallel algorithm development and analysis, and parallel systems and programming languages. Moreover, the optimization of the application of massively parallel architectures to real world problems may provide the impetus for the development of entirely new approaches to applied mathematical modeling. The papers in this volume represent many of the presentations at a workshop whose theme was the simultaneously consideration of the applied mathematical, computer science, and application aspects of parallel scientific computing. We believe that such an interdisciplinary approach is likely to lead to the most rapid possible advances.

We wish to thank the Institute for Mathematics and its Applications of the University of Minnesota and in particular its Director Professor Hans F. Weinberger for their invitation to organize this workshop and for their assistance and gracious hospitality.

Martin H. Schultz
New Haven, Connecticut

PARALLELIZATION OF PHYSICAL SYSTEMS ON THE BBN BUTTERFLY: SOME EXAMPLES

WILLIAM CELMASTER*

Abstract. We will explore how parallelism is exploited in two physical systems which have been analyzed on the BBN Butterfly™. The first of these systems is a molecular model of liquid sulfur hexafluoride. The other is shear-wave propagation in metal, as analyzed using finite element methods. The dynamical equations, in both cases, involve local interactions. However, the systems are permitted to be spacially inhomogeneous. Consequently, both systems can be effectively simulated with MIMD algorithms. These algorithms will be discussed.

The goal, in this talk, is to show how a couple of scientific programs can be organized to take advantage of the MIMD (multiple instruction, multiple data) shared-memory medium-grain architecture offered by the BBN Butterfly. The first of these programs describes the dynamics of interacting sulfur hexafluoride molecules. The other is a finite element method (coFEM) program to simulate shear-wave propagation in metal. The dynamical equations, in both cases, involve local interactions. Spacial inhomogeneities allow these interactions to vary throughout the system. There are many other examples of such dynamical systems, and the methodology to be described will apply in those cases as well.

I. INTRODUCTION

As we move into the era of dusty-deck Cray programs, it is well to remember that the scientific or mathematical problems from which they originated may have been structured very differently from what they are at present. They may, for instance, have been naturally formulated as MIMD, rather than SIMD algorithms. Of course, certain specific library routines such as Gaussian elimination and fast Fourier transforms, have been the subject of a tremendous amount of numerical analysis research. For these "kernel" algorithms, all sorts of useful restructuring has been exploited and the scientist can choose the formulation most appropriate for the available architecture. But in large-scale scientific and engineering projects, there usually can be found substantial portions of code which have not previously been optimized for any particular architecture. The question then becomes, "How easy is it for the the scientist or engineer to obtain efficient performance?"

One of the goals of present and future BBN Butterfly Computer architectures is to facilitate the parallelization of algorithms. There are at least two ways in which these goals are achieved: (1) The MIMD nature of the Butterfly permits the general exploitation of parallelism, and (2) the shared-memory model allows maximum flexibility in data management. The first part of this talk will be a description of how the Butterfly architecture attempts to achieve the goals of providing a MIMD shared memory programming environment. In the second part of the talk we will explore a methodology for parallelizing, in the Butterfly environment, certain kinds of physical problems.

II. BUTTERFLY ARCHITECTURE

It might be ideal to have a number of independent processors whose memories are all interconnected. If there were P such processors, then there would be $O(P^2)$ wires connecting them in all possible pairs, and each slab of memory would require $O(P)$ ports. With present technology, this would be prohibitively expensive. Therefore other connection schemes must be used. BBN has chosen to use a network-switch technology. Noteworthy features of the Butterfly switch are:

- It uses packet-switching techniques developed over the years at BBN.
- Data is transferred serially between processors.
- There are $O(n \log n)$ switching nodes.

*BBN Advanced Computers Inc., 70 Fawcett Street, Cambridge, MA 02238

- There is an $O(n)$ bandwidth.
- The data rate is 32 million bits per second per path.
- The time to access a data item through the switch is about four microseconds.
- The switch can be reconfigured and expanded to allow for anywhere between two and 256 processors to be connected.

In order to explain how the switch works, a simplified version is shown in Figure 1. (For more details see Ref. [1]). Each circle represents a 2×2 crossbar which connects the two ingoing wires to the two outgoing wires. (The actual switch has 4×4 crossbars rather than 2×2 crossbars.) Processors P_1 through P_8 can then each find a path, via the crossbars, to the other seven processors. Figure 1a shows data being sent from P_1 to P_5 and other data being sent from P_5 to P_3. In this case there is no conflict between the messages. For contrast, Figure 1b shows that when data is sent from P_4 to P_6, both paths require a common wire. There is contention between the two messages and only one piece of data gets through. The other must wait and retry. In the larger Butterflies, an extra bank of nodes is introduced in order to provide alternate paths. This helps to reduce contention.

Each processor board has a Motorola MC68020 processor with an MC68881 floating-point coprocessor. On each board, processor and interprocessor memory management is performed with a processor node controller (PNC). Each board comes with one megabyte of memory and is expandable to four megabytes, providing a maximum possible memory of one gigabyte for a fully expanded system. At the moment, program development is done on a host such as a VAX or Sun Microsystems workstation, and programs are downloaded to the Butterfly. C, Fortran and Lisp (both Common Lisp and Scheme) are available. The Butterfly native operating system is called Chrysalis. Application programmers can access the machine parallelism through Chrysalis routines, as well as through a higher-level library of routines known as the Uniform System.

In order to effectively program the Butterfly, the programmer must clearly understand the distinctions between different types of memory. Each processor can reference memory on either its own board or it can access memory, through the switch, on another processor board. The first type of memory reference is called *local* and the second is called *remote.* Although the switch is very fast, remote contention-free references take from three to five times longer than do local references. Faster access to remote data can be obtained by using block transfer operations. The distinction between remote and local memory is handled entirely by the hardware, so the programmer does not have to keep track of where a given word memory is located. However, various Chrysalis and Uniform System routines do allow for specification of which processor board should contain which data. Generally, the programmers choose to be sheltered from such decisions.

Another important memory distinction to be made is that between *private* and *shared* memory. This is a software distinction. If a process wants to share some memory with another process, that memory must explicitly be declared as shared memory. Otherwise, it is private. It usually makes sense that private memory be local to the processor on which the process is active and, indeed, Chrysalis supports that concept. Shared memory, on the other hand, can be anywhere.

On the basis of these facts about memory, the following considerations apply to performance tuning:

- When possible, memory references should be local.
- When remote references must be performed, it is best to use block transfers of data if possible.
- Contention for shared data should be reduced by avoiding situations where several proces-

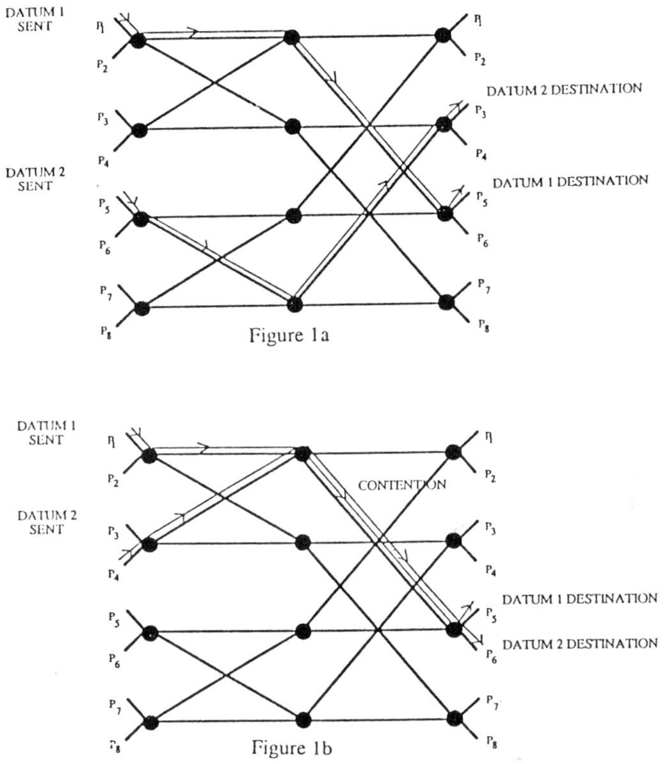

Figure 1: Operation of the Butterfly switch

sors simultaneously try to access memory on a single processor. This contention can be further reduced by scattering related data amongst processors.

III. The Physical Models

The first of the two models to be discussed here is the molecular dynamics simulation developed by G.S. Pawley [2] and C. Baillie with the help of E. Tenenbaum and W. Celmaster. The problem is to simulate the dynamics of SF_6 (sulfur hexafluoride) molecules, whose interatomic interactions are given by the Lenard-Jones potential:

$$U(r) = \frac{a}{r^6} - \frac{b}{r^{12}} \tag{1}$$

The temperature (kinetic energy) is initially quite large, so the molecules jiggle around randomly in the liquid phase. The liquid is gradually cooled, in the simulation, by scaling down the molecular velocities and thus drawing off kinetic energy. When the temperature has finally dropped to a low enough value, one expects to find that the molecules will have arranged themselves in a crystalline structure. This is the solid phase. Even more interesting is the plastic phase, which is intermediate between the liquid and solid phases. In fact, one of the major scientific goals of the present work is to study the liquid-to-plastic transition.

The other physical model to be discussed is the simulation of metal when subjected to stress or strain. The work to be discussed was done by H. Allik, S. Moore, E. O'Neil and E. Tenenbaum [3]. Two related engineering problems were investigated by these people. The first of these-the so-called transient problem-is the analysis of a shear wave in a piece of metal subjected to a time-dependent force. The other problem studied by Allik and collaborators is the static problem, where internal stresses are analyzed in a strained piece of metal which is in equilibrium. In both of these investigations of metal, the elasticity equations were analyzed by the finite element method (referred by the authors as coFEM). Today's talk will present the analysis of the transient problem. (The static analysis was computationally dominated by the linear equation solver, which was based on the method of Gaussian elimination for reasons of robustness. Other solvers are presently under investigation by Allik and Moore, and are expected to result in more efficient parallelizable code.)

Although the two systems-molecular dynamics and shear wave dynamics-seem vastly different, they share some critical features:

- Both are governed by Newton's laws of motion.
- The interaction forces are local.
- The systems are permitted to be spacially inhomogeneous in their local connectivity. More precisely, in the molecular dynamics case, the number of effectively interacting neighbors can vary from one molecule to the next. In the case of stressed metal, the local shape can vary from one position to the next (consider an airplane wing), and some regions may experience plasticity, i.e., nonlinearity.

Many other physical systems share these features. Therefore it is worthwhile to try to identify some common methodology for parallelization.

IV. Methodology

A. Data Storage. Both algorithms involve manipulations on shared arrays which describe the state of the system. The molecular dynamics (MD) arrays can be described as C language structures of the form STATE[MOL]. Each state structure contains, for instance, the x, y and z coordinates of each of the sulfur and fluorine atoms. Examples would be

STATE[MOL]. fluorine.1.x,
STATE[MOL]. fluorine.2.y, etc.

Velocities, accelerations, and several other quantities are also stored. Similarly, the coFEM arrays describe the nodal displacements, velocities, and forces associated with each element, as well as structural information identified with that element.

How should these arrays be stored? There are two apparently orthogonal techniques for storage. The first (block storage) is reminiscent of the style used for caching and out-of-core algorithms, and facilitates remote/local performance tuning. The other technique (scatter storage) is special to the switch architecture and is required for contention reduction. In order to understand these two methods, consider a vector $V = (V_1, \ldots, V_6)$, which is to be shared among three processors, P_1, P_2, and P_3. Suppose that each processor needs to modify some of the entries in V. With block storage, V would be stored in a block of contiguous memory in, say, processor P_1. Then all three processors could block-copy V to local memory, perform the calculations, and block-copy the results back to V. Although the transfer time is decreased through the use of block-copy (as opposed to single-word remote references), there is strong contention for processor P_3, and that will degrade the performance. On the other hand, if the goal was contention reduction, V would be "scattered" among the three processors. For instance, V_1 and V_2 would be stored in P_1, V_3 and V_4 would be stored in P_2, etc. Block-copying would no longer be effective, but contention would be reduced significantly.

In practice, the situation is a great deal more complex. For example, with arrays, one can combine the block storage and scatter storage methods. Rows can be distributed to different processors. For many computations this both reduces contention and allows effective use of block transfers. In fact, this operation of array scattering is so common, that it is done automatically by a single Uniform System instruction, *AllocScatterMatrix.*

This operation was used in both the MD and coFEM programs for storing the state arrays. However, it may not always be possible to store data in such a way that both block storage and scatter storage are accomplished simultaneously for a given access pattern. Studies done on the MD problem indicate that it is best to opt for scatter storage (in fact, as shown in a later section, block transfers improved the MD performance by only about 10 per cent).

B. General Organization of the Calculation.

(1) Molecular Dynamics.

1. From the relative positions of the atoms, the Lenard-Jones potential is used to compute the forces acting on each atom. The only pairs considered are neighbor pairs defined by $|\vec{r}_1 - \vec{r}_2| \leq d$.
2. (b) From the forces on each molecule, the accelerations are computed using $F = ma$, and then the new atomic displacements are computed from those. The fundamental equations used are Beeman's [4] finite-difference solutions to Newton's second law:

$$x_{n+1} = x_n + v_n(\Delta t) + \frac{1}{6}(4a_n - a_{n-1})(\Delta t)^2 \tag{2}$$

and

$$v_{n+1} = v_n + \frac{1}{6}(2a_n + 5a_{n-1} - a_{n-2})(\Delta t) \tag{3}$$

3. Derived quantities, such as pressure and energy, are computed. Adjustments are made based on these quantities, to the positions and velocities. The program is then continued from step 1.

(2) Transient Analysis.

1. From the displacements and velocities of nodal points in the elements, the stress-strain calculation is performed in order to obtain the nodal forces.
2. From the nodal forces, the accelerations are computed and then integrated to find the new displacements and velocities. The finite-difference solutions to Newton's second law are, in this case, the central-difference solutions

$$v_{n+\frac{1}{2}} = v_{n-\frac{1}{2}} + a(\Delta t) \tag{4}$$

and

$$x_{n+1} = x_n + v_{n+\frac{1}{2}}(\Delta t) \tag{5}$$

3. Derived quantities, if any, are computed. Control then returns to step one.

C. Dynamical Load Balancing. Since the state updates are done for each molecule (or element), the natural unit of work is indexed by the molecule (or element). On a serial machine, this would be indicated by a DO loop over molecule (or element) indices:

$$\text{for } (i = 0; i < \max;\ i{+}{+})\ \text{update } (i);$$

On the Butterfly, using the Uniform System, this is converted to

$$GenOnI(update, max);$$

The procedure *GenOnI( )* dynamically allocates tasks such as *update*, to processors. If there are P processors available, the first P iterations (*update*(0), ..., *update*($P-1$)) are farmed out to those processors, and as soon as one of those processors is finished with its task, it obtains the next available task on the "task-queue." It is not important whether all tasks take the same amount of time to complete. If some processors are finished earlier than others, they just grab the next available task and continue. Efficient MIMD computation can therefore be done.

The dynamical allocation should be contrasted with static allocation, where tasks are pre-assigned to processors. In the static case, straggling tasks may eventually force all other processors to be idle for significant amounts of time. This does not occur in the dynamical balancing, provided that the number of tasks far exceeds the number of processors, MAX>> P.

There is a further constraint which must be satisfied for efficient tasking. This granularity constraint is that the task-size must be much larger than the context-switch overhead associated with obtaining a new task. The Uniform System reduces the context-switch overhead by pre-loading all processors with the program template. Thus, rather than MIMD, the style here is STMD - single task, multiple data - which is similar to the SPMD paradigm of Darema-Rogers, Newton and Pfister [5].

Both the molecular dynamics and coFEM programs satisfy the tasking constraints of the above paragraphs. Efficient parallelization is therefore possible. The inhomogeneity of the physical systems would make it difficult to efficiently perform the computations on a SIMD or vector machine, however.

D. Newton's Third Law, Atomic Operations and More Load-Balancing. Newton's third law asserts that if $\vec{F}_{AB}$ represents the force that A exerts on B, then $-\vec{F}_{AB}$ turns out to be the force that B exerts on A. The computational significance of this law is that if you have updated the force contribution to A, due to B, then you do not have to separately compute B's contribution to A.

In the molecular dynamics problem, this observation is implemented by the following: Let MOL be the molecule under consideration (it is the argument of *update( )*). Its force contributions come from its neighbors. A given neighbor, indexed by NAB, contributes to the force only if NAB > MOL. However, whenever $\vec{F}_{\text{MOL,NAB}}$ is computed, NAB's force is updated by $-\vec{F}_{\text{MOL,NAB}}$. Thus half of the computations are saved. It is easy to persuade oneself that, when using this scheme, if the molecular numbering scheme is essentially random, then the task size will tend to decrease as MOL increases. That is because there will be a tendency for fewer neighbors to satisfy the inequality NAB > MOL. As a result, load balancing is even more successful than one might otherwise have guessed. The reason is that, even if there are end-stragglers at the barrier (MOL = MAX), they will generally be of smaller-than-average size, so the straggler penalty is decreased. If the algorithm had been, instead, that force contributions are computed if NAB < MOL (rather than NAB > MOL), the above analysis would lead to the conclusion that straggler penalties would be more severe and that load balancing would be less successful.

Because molecule MOL updates the forces on other molecules, as well as updating its own force, the force update operation must be done *atomically.* That is, in a parallel processing situation, it could happen that two molecules are simultaneously trying to update the force on a third molecule. If the update consists of several machine instructions - i.e., it is not atomic - then the two update efforts will interfere with one another. It is therefore necessary to *lock* the update sequence and to *unlock* it when it is done. This is illustrated in the pseudocode below.

For all neighbors NAB with NAB > MOL {

(1) S = force between MOL and NAB

(2) update a local force, $F \to F = F + S$

(3) LOCK [NAB];
STATE [NAB].force = STATE [NAB].force - S;
UNLOCK [NAB]; }

LOCK[MOL];
STATE[MOL].force=STATE[MOL].force $+F$;
UNLOCK [MOL];

The Uniform System provides simple LOCK and UNLOCK instructions so that the pseudocode implementation is straightforward.

Similar considerations apply to the transient analysis, although Newton's third law appears in a more indirect fashion. Each element shares its nodes with other elements. When the node force is updated by an element, it must also be updated on its neighbor's node. This update occurs atomically just as for the molecular dynamics calculation.

One further point worth mentioning, while on the subject of locks, is the fact that lock operations introduce both contention (for the key) and serial bottlenecks with respect to the locked sections of code. These effects are reduced considerably by using a separate lock for each molecule or element, as indicated in the above pseudocode.

V. Wavefront Sychronization

In both the molecular dynamics and coFEM analyses, each time step involves at least two parts: the force update and the displacement/velocity (d/v) update. The force update in a given spacial region must be completed before the d/v update and vice versa. In the simplest implementation of this sequencing, all molecules (or elements) complete their force update before they all begin their d/v update. This sequencing is known as barrier synchronization, since a "barrier" is erected which is removed only upon completion of the last task in the set. As described earlier, if the number of tasks far exceeds the number of processors, barrier synchronization is fairly efficient. However, there is a better type of synchronization which can be used in physical systems of the type described here. The physical basis for wavefront synchronization is this: in a dynamical system with local interactions, information tends to propagate through the system at a certain speed. Even though a certain region in the system may suddenly start to experience rapid changes, a distant region will not immediately respond to that fact. Thus, if forces have been updated in portion A of the system, but have not yet been updated in portion B, the computation of displacements and velocities can proceed in portion A, even though they are not yet ready to proceed in B. There is no danger of any portion of the system getting too far ahead of any other portion because all portions are connected by an *information wave* (thus the name "wavefront synchronization").

The details of this technique, for the transient analysis, are described in [3]. A task must wait until its nodal displacements are current. When the displacements are current, the force update is computed for each node and then stored atomically. A tally is kept, at each node, to indicate whether all of its elements have updated their force contributions to that node. If the element-task in question is the last one to update a particular node, n, then it computes the new velocities and displacements for that node. Otherwise, the task exists.

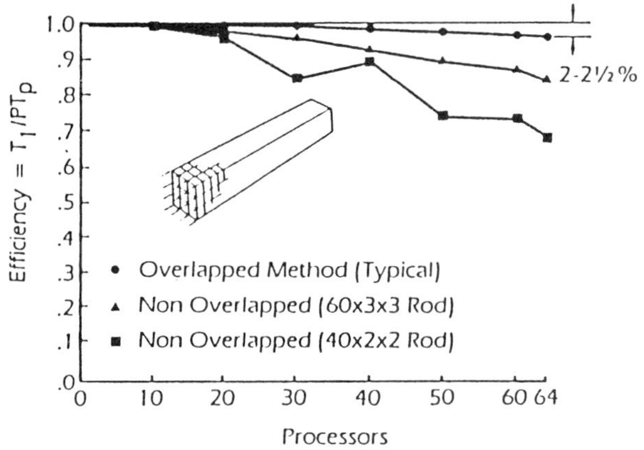

Figure 2: Comparison of Transient Algorithms for Rod Problem

VI. Results

Figure 2, taken from [3], shows the results of running the transient coFEM program on the Butterfly. The absolute time (not shown) is about 30 percent slower, on one Butterfly node, than on a VAX 11/785. Three separate curves are shown. The "overlapped method" refers to wavefront synchronization as opposed to barrier synchronization (not overlapped). The rods are divided into different numbers of elements. Contrast a $60 \times 3 \times 3$ rod to a $40 \times 2 \times 2$ rod. In the latter case there are fewer tasks and one loses efficiency from barrier synchronization. The jagged appearance of the curve is indicative of straggler degradation. The case of wavefront synchronization is clearly more efficient and, in fact, approaches perfect parallelization.

Figure 3 shows the results of running the molecular dynamics program on the Butterfly configured with from 1 through 30 nodes. The analyses were all done with barrier synchronization but would have been slightly improved with wavefront synchronization. In one case, we used an algorithm where none of the remote references utilize the block transfers. For 30 processors we obtain 83% efficiency. In the other case, block transfers were used for some of the remote references. Actually, some unnecessary data was transferred-data which happened to be interspersed throughout the blocks that were needed. Thus this block transfer technique could have been tuned to further enhance the performance. The present implementation yields 93% efficiency for 30 processors.

The main significance of these results is this: excellent performance can be obtained even if none of the data is referenced using block-transfers. Thus, there is no need for the programmer to expend energy in doing block storage. The results suggest that it may be easier to develop algorithms for the Butterfly than it would be for a vector machine, whose performance critically depends on the patterns of data reference.

VII. Summary

For physical systems interacting via local dynamics, a certain programming methodology has been prescribed for the Butterfly. The most important points are:

- Tasks should be indexed by local *elements* such as the molecule, or the finite element.
- State information should be scattered. If block transfers are desired, each element's state should be stored in a contiguous block.
- Tasks are dynamically allocated to processors, using the Uniform System routine *GenOnI*( ) or a variation thereof.
- Force updates should be done atomically by all of those elements that are influenced by a particular force.
- Wavefront synchronization allows disjoint regimes to time-step independently, and thus reduces the effect of straggling tasks.

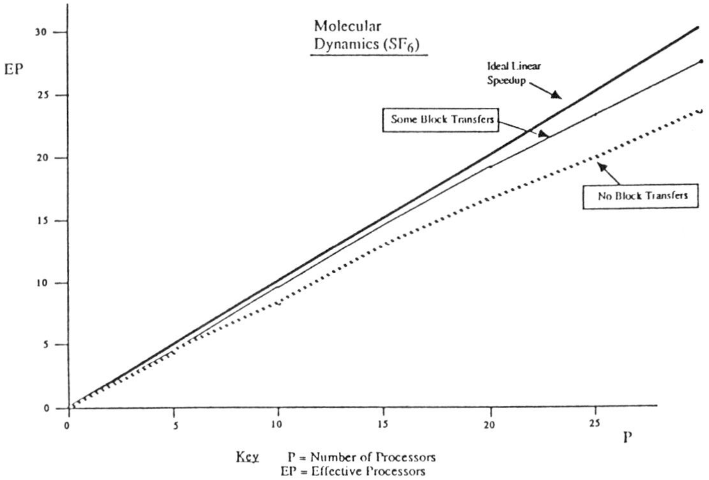

Fig 3: Results of Parallelizing Molecular Dynamics and Transient Analysis

The results, for molecular and transient analysis, as shown in Figures 2 and 3, show that nearly linear speedup (i.e., complete parallelization) is possible through 64 or more processors. Furthermore, and importantly, excellent results can be obtained even if block transfer methods are avoided.

REFERENCES

[1] *BBN*, Technical Report, # 6148 (1986).

[2] G.S. Pawley and G.W. Thomas, Phys. Rev. Lett, 48, 410 (1982).

[3] H. Allik, S. Moore, E. O'Neil and E. Tenenbaum, *Butterfly Finite Element Software*, BBN Technical Memorandum, No. NL-206 (1986).

[4] D. Beeman, J. Comput. Phys, 20, 130 (1976).

[5] F. Darema-Rogers, V.A. Norton and G.F. Pfister, IBM Research Report RC 11552, (#51726) (1985).

SOLUTION OF LARGE SCALE ENGINEERING PROBLEMS USING A LOOSELY COUPLED ARRAY OF PROCESSORS

E. Clementi, D. Logan, V. Sonnad
Scientific and Engineering Computation
Mail Stop 428
IBM, Neighborhood Road
Kingston, NY 12401

I. Introduction :

Our interest in parallel processing systems has been motivated by the need to solve large scale computational problems occuring in theoretical chemistry, biophysics, and more recently in engineering. We have been experimenting with a loosely coupled system, that uses a master processor to control a group of independent slave processors. This architecture was designed to meet the following objectives : the initial system

1) should be classifiable as a "supercomputer",
2) be easily extended to higher computing speeds,
3) be flexible and fault tolerant,
4) permit easy migration of large codes to parallel execution,
5) be relatively inexpensive.

To achieve this goal, an experimental system at IBM Kingston has been configured with a variety of IBM mainframes as masters (43xx, 308x, and 3090 systems), and 20 powerful attached processors as slaves. The latter are either FPS-164s or FPS-264s, manufactured by Floating Point Systems, chosen because of their high processing speeds and the inclusion of 64-bit floating point hardware as a standard feature.

Large codes for scientific and engineering computations are usually written in Fortran. If we wish to execute such codes in parallel, then we need extensions to the Fortran language which permit the expression of various parallel constructs, such as fork-and-join, task synchronization, and mutual exclusion of critical sections (with locks). This approach has been implemented through the developement of a precompiler which processes such directives and expands them into standard systems communications software prior to compilation. In the same spirit, we strive to minimize the additional coding necessary for parallel execution, and try to retain as much as possible, the structure of the original sequential code. On the whole, we consider our approach to parallel processing to be "off the shelf"; this has been an important factor in the rapid and cost effective development of our system.

A detailed description of the basic system is given in Section II; operating system considerations are given in Section III. In Section IV we describe extensions to the system hardware that have enhanced the basic architecture. In section V we discuss strategies for parallel computation, and in section VI present results on several algorithms and practical applications. Section VII ends this paper with a few concluding remarks.

II. Basic Configuration of lCAP :

There are at present two parallel processing systems functioning in our laboratory; both share the same fundamental architecture of a host with attached processors. The original configuration did not have a direct path for transmission of data between processors, and hence the system was termed lCAP, (for "loosely Coupled Array of Processors"). The first of these systems, called lCAP-1, is hosted either by an IBM 3081 or 4381 and attaches to 10 FPS-164 array processors, (AP's). The second and more powerful system, called lCAP-2, employs as host a dyadic IBM 3084 and 10 FPS-264's as slaves. The two systems are architecturally very similar, and differ mainly in the operating system used by each. For the purpose of brevity we will confine ourselves to a description of of the lCAP-1 system as it was operating in the second half of 1986. We emphasize that both systems are experimental, and evolve continuously in both hardware and software.

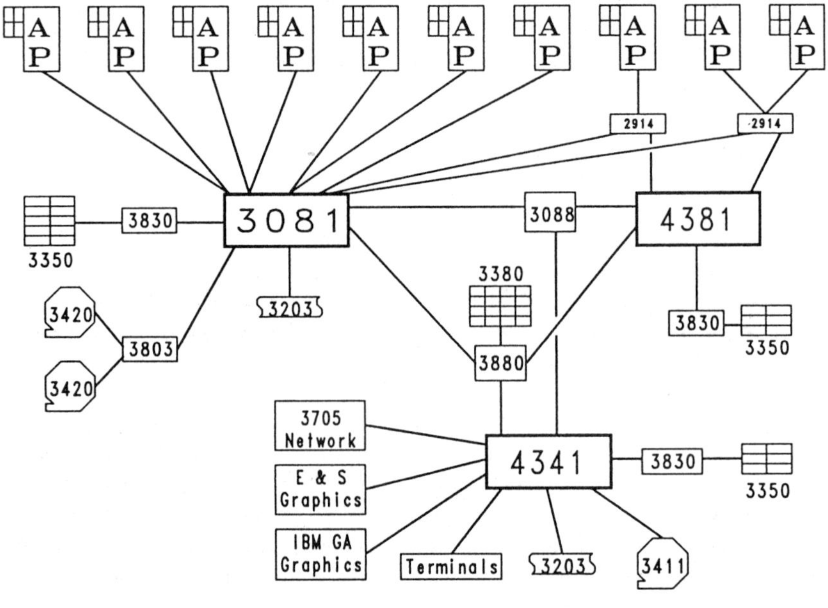

Figure 1: Schematic layout of the basic lCAP-1 system.

lCAP-1 is structured so that six FPS-164's are connected to an IBM 3081 host, and the remaining four are attached so that they can be switched between an IBM 3081 host or the IBM 4381 host. This gives us the flexibility of a "production" system with six AP's and a "development" system with four AP's. The AP's attach to the hosts through IBM 3 Mbyte/sec channels. A schematic diagram of the configuration appears in Figure 1. Each AP contains an independent CPU and its own memory and disk drives. The CPU on the

FPS-164 is capable of 11 million floating point operation per second (peak performance). Each of our FPS-164's has 8 Mbytes of random access memory, and is connected to four 135 Mbyte disks, for a total of 5.4 Gigabytes. In addition there are banks of IBM 3350 and IBM 3380 disks accessible to the host computers, totalling about 25 Gigabytes of disk storage. Tape drives, printers, and communication network interfaces complete our configuration.

To complete the description of lCAP-1, we mention that each of the AP's has been equipped with two FPS-164/MAX boards; these are special purpose boards supplied by FPS, each of which contains two adders and multipliers, and can speed up matrix operations to a peak of 22 Mflops/board. This has upgraded our peak performance from 110 to 550 Mflops. Ultimately our system could grow to 3410 Mflops peak capability, using 15 boards per processor. However it is clearly desirable to first explore the gains that one can realistically obtain with only a few 164/MAX boards per AP, so we have settled at a peak performance of 550 Mflops.

III. Operating System Considerations :

The lCAP-1 system, hosted by either an IBM 3081 or 4381, runs under the IBM Virtual Machines/ System Product (VM/SP) operating system (Ref. 1). For the AP's, we use the systems software provided by Floating Point Systems (Ref. 2). We have not found it necessary to modify either sets of software in order to run our applications in parallel. VM/SP is an operating system in which jobs run on virtual machines (VM) created by the system; these VM's simulate real computing systems. The standard software provided by Floating Point Systems allows only one AP to be attached to a VM. Since we need more than one AP for a task running in parallel, our solution has been to introduce "slave" VM's to handle the extra AP's we need. The communication between processors is transparent to the user since the details are handled by the precompiler.

The lCAP-2 system utilizes as host, a dyadic IBM 3084 that runs under the IBM Multiple Virtual Systems (MVS) operating system . Connection to each individual AP is effected within the MVS operating system through the creation of subtasks; communication between the subtasks and the attached processors takes place as in lCAP-1. However, in the MVS environment, system software permits the overlap of virtual memories of master and slave subtasks such that they may share common memory. The cost of data transfer between host and slave subtasks is thus significantly reduced in the MVS system.

Regardless of the differences in communications in the two operating systems, it is important to note that programs written for one system are easily migrated to the other, since the AP software remains the same, and the precompiler handles the remaining details.

IV. Extended Architecture :

In our initial configuration, each AP could communicate only with a slave VM or subtask; AP to AP communication was achieved indirectly by having an AP communicate to the host, which then communicated to the second AP. However, communication overhead is an important consideration in some applications, so the lack of direct AP to AP communication was a serious limitation. To address this problem, the system has been extended using three independent but complementary approaches.

The first is the incorporation of 5 locally shared bulk memories that link the 10 AP's. Each memory is 32 Mbytes in size and is multiply connected to four attached processors, (Fig. 2). This configuration is that of a ring with multiple connections per AP allowing added flexibility and fault tolerance. Each memory is capable of sustained data transfer rates of 44 Mbyte/sec when attached to the FPS-164. Communication between the AP's and bulk memory is carried out by Fortran callable routines, (Ref 3).

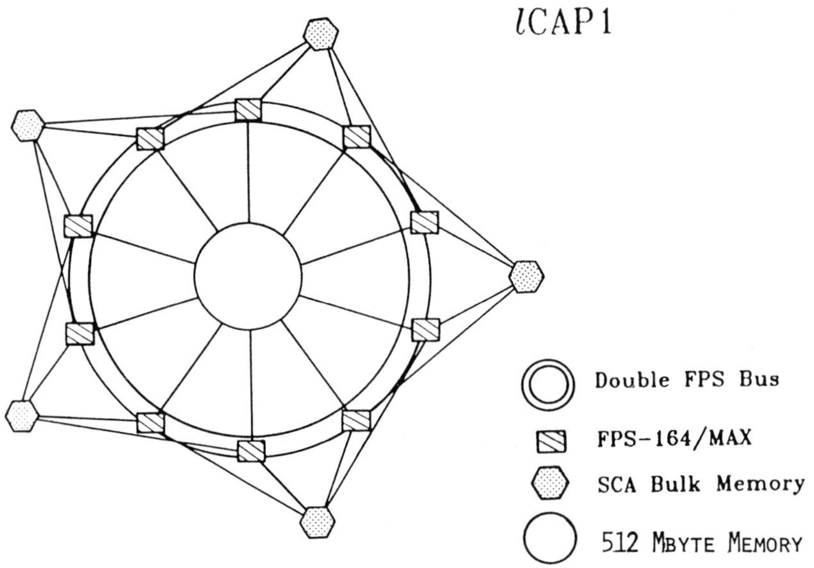

Figure 2: Extensions to ICAP-1 architecture.

The second feature, also indicated in Figure 2, is the addition of two fast buses that link all of the APs in a ring configuration. This bus, designed and built by FPS, is capable of transferring data at rates of 22 Mbytes/sec between AP's.

A third major addition is the installation of a globally shared memory of 512 Mbyte capacity (again shown in Fig. 2). Each one of these global memories can be connected to 12 processors. Memory interleaving permits up to three processors to talk to the memory simultaneously, allowing a maximum data transfer rate of 132 Mbytes/second. This modification allows both ICAP-1 and ICAP-2 to be optionally configured as a tightly coupled array of processors.

We therefore have four independent, non-conflicting paths for data transfer between AP's : IBM channels for slave-host-slave communication, the bus for message passing and broadcasting between two processors, the ring of locally shared memories for nearest neighbour communication, and the central memory for asynchronous communication between any two AP's within a system. This will allow us to experiment and implement algorithms that can make effective use of one or more of these data paths.

A third system termed ICAP-3, is essentially a coupling of ICAP-1 and ICAP-2, using an IBM-3090 with high performance vector features as a master; the interconnect scheme is shown in Fig. 3. The goal is to achieve a system with a high level of flexibility, and the ability to achieve high sustained rates of computation on problems that are not easily solved by today's supercomputers.

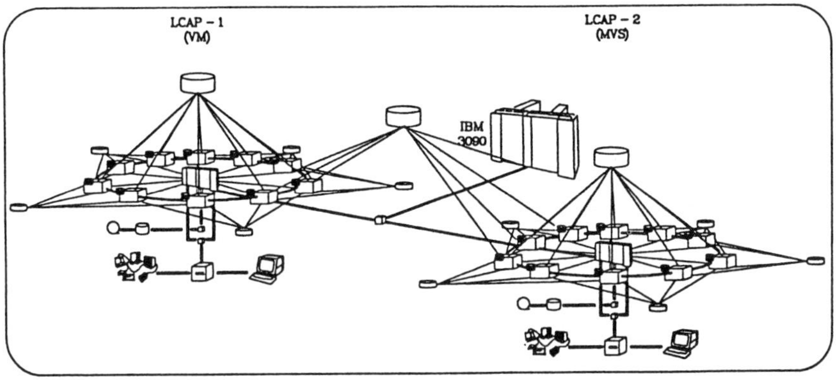

Figure 3: Configuration of ICAP-3 system.

V. Strategies for Parallel Processing In Engineering :

The initial applications on ICAP were devoted to areas of computational chemistry and biophysics (Ref. 4). The considerable success of this architecture in speeding up solutions of such problems, has encouraged exploration in applying this architecture to other fields. In this paper we explore some approaches for parallel implementation of engineering problems.

Applications programs in engineering can be very broadly grouped into a few major areas: fluid mechanics, heat transfer, structural mechanics, and electro-magnetism, (corresponding roughly to departments within an engineering school), and the field of optimization which is interwoven into almost all of the above areas. The list of applications within each field however is almost endless, and any attempt to approach parallel computation on an application by application basis is an unending task. For the purpose of devising a strategy for parallel computation within engineering, it is much more constructive to examine the underlying methods that are used in various applications. The most commonly used methods for the solution of problems in engineering are the following :

Finite differences : This method is further broken down into the categories of explicit and implicit methods. The major computational component of explicit methods, is the calculation of new values from old by substitution into algebraic expressions; implicit methods require the solution of sparse systems of coupled algebraic equations. It is useful to have another category of semi-implicit methods, where the problem is formulated as an implicit method, but is solved by methods such as ADI or SOR, which decouple large coupled sets of equations into smaller, more manageable problems.

Finite elements: This method has historically been linked to structural engineering, but is now becoming widely accepted as a general method for the solution of partial differential equations. While it is possible to use explicit schemes with this method, implicit methods are the rule, and for large problems, the solution of coupled systems of large sparse equations, and the calculation of the eigenvalues and eigenvectors of sparse matrices form the major computing component of this method.

Spectral and Pseudo-Spectral Methods : The basic concept used in these methods is to transform the problem from the physical domain into the spectral domain using typically Fourier or Chebyshev transforms. The problem is then addressed in the spectral domain, and an inverse transform is performed to get the answer in the physical domain. For linear problems, one forward and reverse transformation is sufficient, and the major computing component is the series of operations performed in the spectral domain. For non-linear problems it is necessary to transform back and forth between the physical and spectral domains at every time step, and the major computational component is the transformation operation.

Boundary Elements : This method has an advantage over other discretization methods, because it operates on integral equations, and hence reduces the dimension of the problem by one: e.g., a three-dimensional body needs to be discretized only on its surface and not over the entire volume. This can be a distinct advantage when the solution is required only on the boundary. This method requires the solution of a system of equations that is dense (and usually nonsymmetric); equation solvers that apply to sparse systems are inappropriate. LU decomposition is a suitable approach, because there is no problem of fill; furthermore, the equations are normally well conditioned , and pivoting is not necessary.

VI. Parallel Algorithms On ICAP :

It is seen from the discussions in the previous section that there is a set of core algorithms that form major computational components of several methods used in engineering analysis. By focusing on the parallel implementation of these algorithms, it is possible to address applications within several different branches of engineering. We discuss below, the parallel implementation of some of the more important algorithms, and describe their performance on the ICAP system.

1) Householder Reduction of a Matrix to Tridiagonal or Hessenberg form :

Because of its general importance in engineering and applied science and its high operation count, (order n**3 for an n x n matrix), the eigenvalue problem has been the subject of repeated efforts over the past decade to determine efficient parallel algorithms. Most of these have been directed towards the Jacobi method, where the element of concurrent computation is the zeroing of elements or blocks that do not share any common indices. Given a suitable interconnect scheme of parallel processors, and ingenious scheduling algorithms, the parallel version of this method has often given near linear speedup.

However, success in this regard is tempered by the fact that the Jacobi algorithm is usually inferior to the method of reducing the system to tridiagonal form, (or Hessenberg form if unsymmetric), followed by the iterative QR factorization and RQ updates. In the following we consider a parallel version of the reduction phase of this algorithm to tridiagonal or Hessenberg form, and its performance on the ICAP multiprocessor system (Ref. 5). Parallel methods for determining the eigenvalues of tridiagonal matrices are described in several articles in Ref. 6

The fundamental algorithm for Householder reduction, for a matrix of size n**2, can be represented as, (Ref 7):

```
   DO 1 i= 1,n
1  A = H(i) * A * H(i)
```

where H is the Householder transformation and is utilized in its outer product form

H = I - w * w(t) ;   (w(t) is the transpose of w).

H(i) is the transform that reduces the ith column to tridiagonal form and is constructed in the usual manner from the last n-i elements of the appropriate column of A, following its (i-1)th update. Parallelism in this procedure is limited in that the output of stage i is the input for stage i+1. However the basic step written as

A(i+1) = (I - w * w(t)) * A(i) * (I - w * w(t))

admits a high degree of parallelism if it is executed in the following three steps:

1. B = (I - w * w(t)) * A(i)

2. c = B * w

3. A(i+1) = B - c * w(t)

Considering a system with p parallel processors, with the algorithm at stage i, these steps may be executed in parallel as follows : We distribute the matrix A(i) equally between processors on a column basis such that the first processor has the first block of (n-i)/p columns, the second, the corresponding second (n-i)/p columns, and so on. Then given that each has the appropriate Householder vector w, each processor can complete step 1 in parallel on its portion of the columns. The second step needs the matrix B of which each processor has an equal portion following the first step. However each processor can compute its contribution to the vector c. Following that, we require a barrier synchronization between all, and the sum of all components to determine c. Thus step two, is simply the parallel implementation of n-1 parallel dot products. Lastly the third step is again performed in parallel, with each processor updating its columns of A(i+1) with a rank one matrix, c *w(t). Following the last step, the next Householder vector is computed by the processor holding the i+1 th column, (always the first processor), and is broadcast to the remaining processors for beginning the next reduction stage.

Load balancing and efficient data movement is achieved by always segmenting the columns evenly between processors. It may be seen that the locally updated columns from the preceding stage can be used immediately for the next stage. At most one column from the computational right neighbor may be required. This minimization of data movement is essential, as the computational operation count for each stage is only of the order of (n-1)**2. Synchronization is again required for these last vector transfers.

A Fortran version of this parallel algorithm was written and benchmarked on the ICAP system. Synchronization and data transfers were implemented using the shared bulk memories described previously. For systems of order 800*800 and larger, the efficiency was slightly above 80% for 6 processors (see Fig. 4); this performance is expected to improve on going to larger matrices.

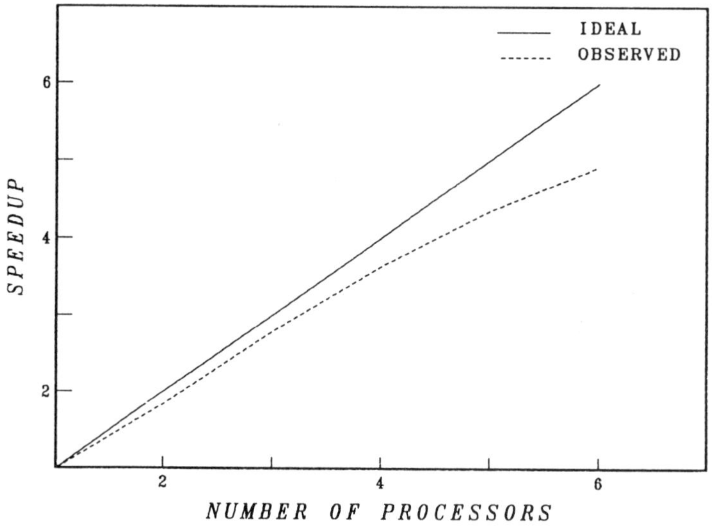

Figure 4: Householder reduction ; (800 * 800 matrix).

Extending the algorithm to problem sizes that are greater than local memory has not as yet been explored. However the following comments may be made . If the problem size n, is such that n*n/p fits in local memory, then as emphasized previously, only one boundary column at most needs to be transferred between adjacent processors at each reduction cycle. If local memory is insufficient, then at the beginning of the algorithm, it will be necessary to write out block columns to make room for new block columns to be updated during each cycle. However the extra penalty in data movement is offset to a large degree, by the much larger parallel computational grain size between such transfers, and should not degrade the performance significantly. Furthermore, as the reduction proceeds to a certain point, the remainder of the reduction becomes entirely in core again, with all the advantages of small data transfers just described. Thus the overall parallel strategy seems to be well adapted to problems of any size beyond some threshold magnitude.

Although the computation of the full transformation matrix was not included in the algorithm described above, the calculation of its transpose can be incorporated in a similarly efficient parallel manner. That is, we parallelize again on a column basis; at any given stage i, the update to this matrix is H(i + 1) * T , where T is the accumulated product of previous Householder transformations. The segmentation of T, and the boundary column data movement between adjacent processors, follows the identical strategy as the matrix A that we are reducing.

Lastly it should be mentioned that a simple QR factorization of a general matrix as contrasted with the more complex Householder reduction described above, can be performed with exactly the same strategy used here. Indeed, the absence of a right application of a Householder transformation removes one synchronization that is needed in the latter. Initial measurements of the parallel QR factorization performed in this manner, indicate almost linear speedup for similarly large sysyems.

2) Block LU Decomposition :

The parallel factorization of a system of equations is done on a block by block basis, but the algorithm is identical to the familiar element by element approach. The operation count is slightly higher in this formulation, (approximately 2% for a 500*500 system), but the execution rate is much higher in block form as each operation now consists of matrix multiplications which are ideally suited to pipelined architectures. Indeed we find that on the FPS-x64 computers the sequential algorithm of choice is the block form.

The important characteristic of this algorithm is that it allows processors to dynamically schedule tasks, without undue considerations of communication that arise in static scheduling of tasks. This attribute enhances the load balancing of the system, and allows processors of differing computational power to work cohesively on the same problem. The essential operation in permitting dynamic job assignment, is the lock mechanism which allows mutual exclusion in critical sections. This is implemented on the ICAP system with the use of shared memory.

Parallelism in block LU factorization is readily apparent and well known. Thus, once any diagonal block has been inverted, all blocks immediately below this block column may be updated to L blocks in parallel. Similarly once an L block has been computed, all blocks along the appropriate block column may be updated in parallel. The only subtlety arises in devising the order in which to process these tasks.

In this regard, the essential feature is to eliminate as much as possible any bottlenecks that arise from data dependencies between updates to blocks. Thus we schedule the update of any block required at a later stage of the factorization, as early as possible; this minimizes the probability that processor(s) have to wait for completion of updates on that block. In this regard we may write the block factorization algorithm for x blocks as:

```
      do 1 k = 1,x
      if ( k.eq.1 ) A(k,k) = A(k,k)**(-1)
         i = k
         do 2 j = k+1,x
2        A(j,k) = A(j,k)*A(k,k)

            do 3 i = k+1,x
            do 3 j = k+1,x
            A(j,i) = A(j,i) - A(j,k)*A(k,i)
            if( i.eq.j .and. j.eq.k+1 ) then
               A(k+1,k+1) =A(k+1,k+1)**(-1)
            endif
3        continue
1     continue
```

Here symbol A(m,n) represents a block element in the mth block row and and nth block column. It may be seen, that the only bottleneck is the computation of the first block inverse in position (1,1). Thereafter we schedule tasks proceeding down the first column to compute the L factors. After that we schedule across the rows starting at the top and working downwards. In addition, the very first job going across the columns following its update is inverted immediately, so that it is ready as soon as possible for the next sweep. The block elements are stored in block contiguous form on bulk memory, and are read and restored to this memory per task. As stressed previously, these tasks are allocated to processors on a self scheduled basis. This is achieved by storing in shared memory a three word counter whose elements are the current indices in the algorithm described above. This 3 element array is accessed in mutual exclusion by the processors, who copy the indices corresponding to the present task into memory, and update the shared memory values according to the rules imbedded in the algorithm. They then release the lock on this critical section for other processors to access. In this manner load balancing is achieved dynamically. It is also worth stressing that the programming of this method is very simple compared to that involved in static allocation schemes.

To make the algorithm robust, it is only necessary to ensure that a processor about to update a given block, be able to ascertain whether that block and any other blocks required, are updated to the appropriate level before proceeding. This is accomplished by storing in shared bulk memory a status matrix, that has a one to one correspondence with the block data matrix. Any block that has been updated to level k (see algorithm), requires that it have its corresponding status set to value k. In this manner a processor on scheduling a task first checks the status of the blocks required before proceeding, (and waits if they are not ready).

The forward solution and back substitution were parallelized in a manner entirely analogous to the factorization. Indeed, the forward solution phase is begun while the last processors are completing the factorization stages, as the upper left portion of the LU blocks are complete. The block backward solve however has a sequential initial job, (the solve for the lowest (n,n) block), before the other parallel tasks can proceed.

This algorithm was coded in Fortran for the ICAP system with results similar to those obtained for the Householder reduction; the efficiency was about 85% using 6 processors for problems of size 800 * 800,

(Fig. 5). There is no penalty at all for problems that would normally be out of core, (as long as it fits on the bulk memory), since the computation is done on a block basis.

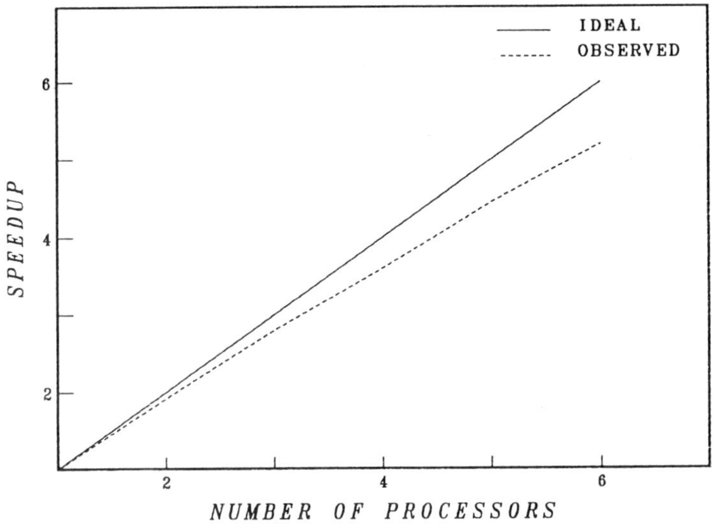

Figure 5: Self-scheduled LU factorization ; (800 * 800 matrix).

3) Two-Dimensional Fast Fourier Transform :

A 2-D FFT can be broken up into a series of one dimensional FFT's (1-D FFT) of length NY (column-wise), followed by a series of 1-D FFT's of length NX (row-wise). The one dimensional FFTs may be partitioned between processors in an efficient and load balanced manner. It is the implicit matrix transpose between column and row FFTs that introduces communications overhead in the total calculations. This transpose can be most easily performed by writing the matrix in blocked form; the row-wise FFT can be performed by reading the blocked data by rows, and then by replacing the blocks with the transformed values. The blocked data is read back by columns and a similar operation is performed on them. Details of this method are given in Ref. 8.

The above method, employing global shared memory was used to produce results shown in Figure 6. Specifically, for a 512 by 512 case the efficiency reached 90% of ideal speedup for a single 2-D FFT; (a problem of size 512x512 is very typical in two-dimensional simulations, and in the absence of symmetries, requires storing matrices of size 512x512 for each of the unknowns in the equations that are being solved). It is observed that the speedup is less for smaller matrices as the computation time becomes small, whereas the transmission time decreases at a slower rate. Although not shown here, the results for the same problem with data transfer through the host had significantly less speedup ; (in fact almost no improvement in elapsed time was observed with additional AP's).

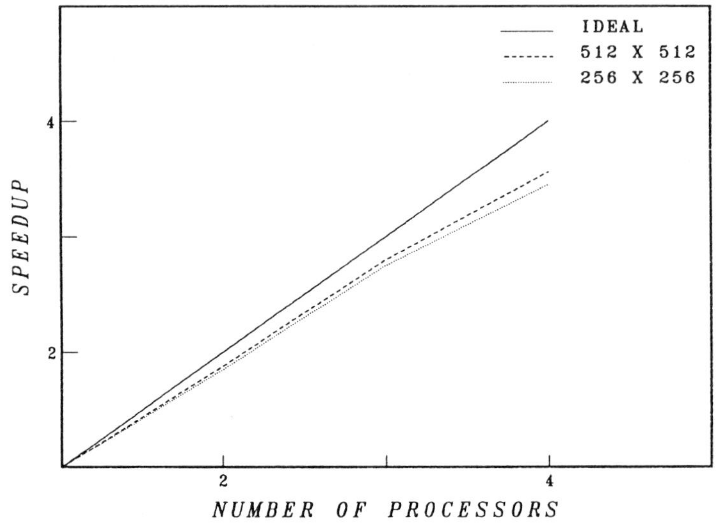

Figure 6: 2-Dimensional Fast Fourier Transform.

4) Preconditioned Conjugate Gradient Methods :

The use of finite elements or implicit finite difference schemes to analyze problems requires the solution of large, sparse systems of linear equations. In the finite element field, it is customary to use direct methods such as Cholesky decomposition, or LU factorization to solve such systems of equations. This is a very difficult method to implement in parallel, (the approach described above for dense matrices, is much less efficient for sparse matrices), and also uses large amounts of storage because the LU factors are much more populated than the original matrix. Instead of using direct methods, we have chosen to work with conjugate gradient methods because they are much more amenable to parallel processing, and have the advantage over direct methods in that they preserve the sparsity of the matrix.

The method of conjugate gradients (CG), works by successively refining an initial guess until the solution is achieved, (Ref 9). For a symmetric, positive definite matrix of size NxN, CG reaches the solution in at most N steps in exact arithmetic; it is thus different from other iterative methods and can be considered as a direct method. When used as such, it is much more expensive than Cholesky factorization for dense matrices; however, for large, sparse matrices it is competitive in operation count and requires much less storage. The use of preconditioning, (Ref. 10), can greatly reduce the number of iterations required for CG to converge to a satisfactory answer, and the method then has an advantage over direct methods in both storage and operation counts. Incomplete Cholesky preconditioning, is generally regarded as the the most effective preconditioner for symmetric, positive definite matrices, but it is also very difficult to implement in parallel. We have therefore experimented with two different kinds of preconditioners that are more amenable to parallelism, and discuss below the implementation of these two approaches on the ICAP system.

a) Diagonal Preconditioning: This is the simplest form of preconditioning, where the inverse of the diagonal of a matrix is used as an approximate inverse of the matrix. For many problems, this simple approach can often reduce the overall iterations by a factor of two or three, at very little additional expense per iteration ; it is also readily implemented in parallel. The major computational steps of the conjugate gradient method consist of a sparse matrix-vector multiplication, scalar products of vectors, and the addition of a vector to scalar multiples of another. These steps are carried out using the globally shared memory to pass data between processors.

This method has been implemented on ICAP (Ref. 11), and the results are shown in Fig. 7. One interesting result is that it is possible to obtain efficiencies greater than 100% . The reason for this is that when the problem was solved on a single processor, it was too large to fit into core, and hence required transmitting large amounts of data to and from the bulk memory at each iteration. When the problem was divided among processors, it could be executed in core, thus reducing data transmission with the bulk memory. This example illustrates that parallel processing can benefit both from additional processors and additional memory.

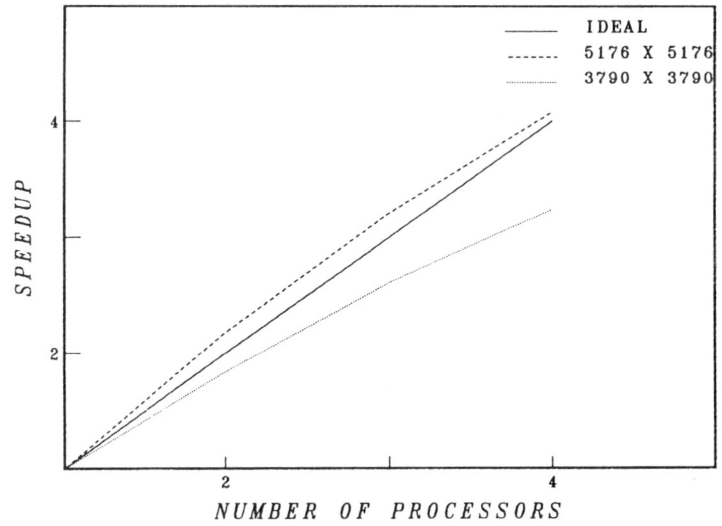

Figure 7: Conjugate gradients with diagonal preconditioning.

b) Element by Element Preconditioning: This approach was originally proposed as a means of reducing the costs associated with the solution of large, transient heat conduction problems. It was later combined with an iterative scheme where the element by element factorization supplied the approximate inverse for preconditioning, (Ref 12).

The essential concept used in this approach, is that an approximate inverse for the global stiffness matrix can be obtained from an expression that contains the products of the inverses of the individual stiffness matrices. In order to preserve symmetry and positive definiteness, it is necessary to perform a two pass decomposition to get an approximate inverse; details are given in Ref. 12.

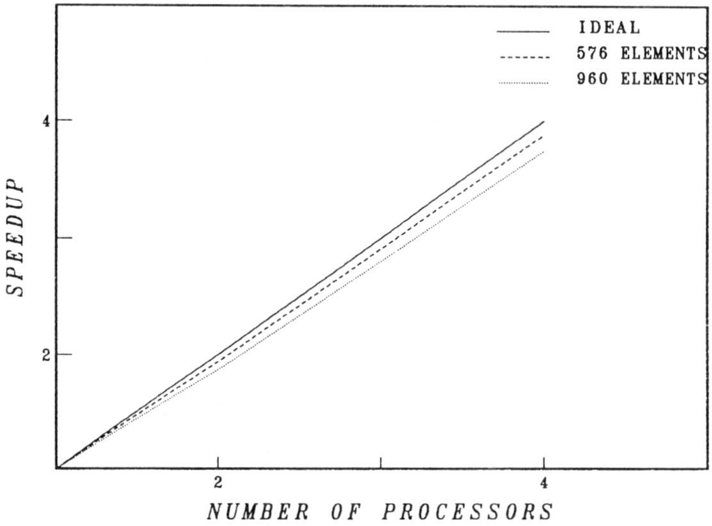

Figure 8: Conjugate gradients with element-by-element preconditioning.

This method has been implemented on ICAP, (Ref 13), and results shown in Fig. 8. The efficiency of this method is among the highest of any of the methods considered so far, and furthermore, does not vary significantly with the size of the problem being solved. This appears to be a very promising method for implementation on a large number of processors.

5) Simulation of Shallow Water Behavior:

This application is presented as an example of an explicit finite difference method using both the ring and master-slave configurations. We are presently engaged in studying, (in collaboration with the Woods Hole Oceanographic Institute), the behavior of a body of shallow water subjected to wind and tidal forces. The problem is formulated in terms of coupled, nonlinear partial differential equations, with the boundary conditions specified along the coast line and at the inlets and outlets.

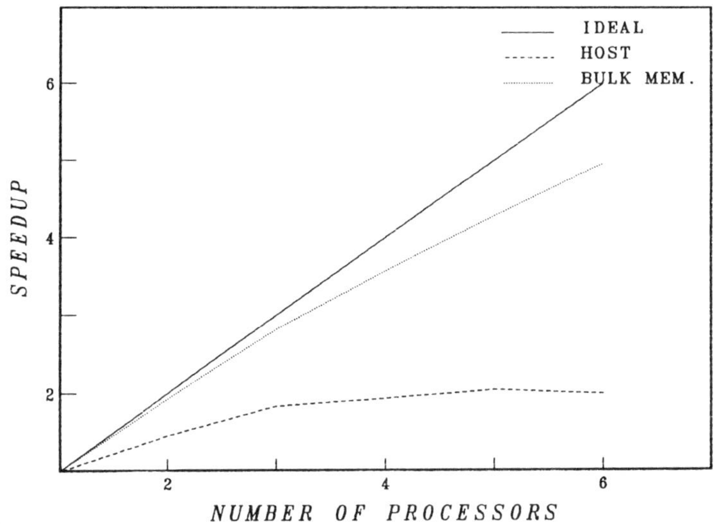

Figure 9: Comparison of performance using two modes of data transfer.

To perform these calculations in parallel, each slave is given an equal portion of the grid to solve at each time step. At the end of a time step, each processor must exchange boundary information with its neighbors. When this information is exchanged using a master-slave configuration, there is a significant communication overhead, and this severely affects the speed-up that can be obtained using several processors, (shown in Fig. 9, bottom). However, when the same calculations are carried out using locally shared memories for communications, the performance is vastly improved (Fig. 9, top). Details of this computation are discussed in Ref. 14.

These results suggest that explicit finite difference schemes may be successfully implemented in parallel, using a nearest neighbor ring architecture. Multiple, locally shared memories permit simultaneous interprocessor communication.

VII. Concluding Remarks :

The ICAP system seems to be well adapted to the efficient implementation of a large and diverse collection of scientific and engineering problems. The extensions to the system, discussed in a previous section, would seem to make this parallel processing system a truly general purpose machine. Future experiments will be needed to verify this conclusion.

The need for algorithm design is critical for parallel processing systems, and we are interested in developing algorithms that map efficiently on to the topology of ICAP-1 and ICAP-2. To this end we have established joint efforts with a number of centers (Yale, Cornell, and RPI), in developing software for such algorithms as matrix diagonalizations, solution of systems of algebraic equations, and FFT's. It must be mentioned that a similar system has been developed at the IBM Scientific Center at Rome, Italy. Close cooperation between our two groups is expected to further our understanding of parallel applications and algorithms.

References

1. Virtual Machine/System Product System Programmer's Guide, Third Edition (Publication No. SC19-6203-2), International Business Machines Corp. (August, 1983).

2. FPS-164 Operating System Manual, Vols. 1-3 (Publication No. 860-7491-000B), Floating Point Systems, Inc. (January, 1983).

3. Shared Bulk Memory System Interface Software Manual, Version 1, August 1985, Publication of Scientific Computing Associates, New Haven, Connecticut.

4. E. Clementi, G. Corongiu, J. H. Detrich, H. Khanmohammadbaigi, S. Chin, L. Domingo, A. Laaksonen and H. L. Nguyen, "Parallelism in Computational Chemistry: Applications in Quantum and Statistical Mechanics", in Structure and Motion: Membranes, Nucleic Acids and Proteins, E. Clementi, G. Corongiu, M. H. Sarma and R. H. Sarma, Eds., Adenine Press, Guilderland, N.Y., 1984.

5. D. Logan, "The Parallel Eigenvalue Problem on the ICAP Multiprocessor System ", KGN100, IBM Kingston ,1986.

6. Proceedings of the Second SIAM Conference on Parallel Processing for Scientific Computing, held at Norfolk, Virginia, Nov 18-21, 1985

7. G. Golub and C. F. VanLoan, Matrix Computations, Johns Hopkins University Press, Baltimore, Maryland, 1983.

8. Z. Christidis, V. Sonnad and D. Logan, "Parallel Implementation of a 2-D Fast Fourier Transform on a Loosely Coupled Array of Processors", KGN68, IBM Kingston, NY,1986.

9. M. Hestenes and E. Stiefel, "Method of Conjugate Gradients for Solving Linear Systems", Journal of Research of the National Bureau of Standards , V49, December 1952.

10. J. A. Meijrink and H. A. Van der Vorst, " An Iterative Solution Method for Linear Systems of which the Coefficient Matrix is a Symmetric M Matrix", Mathematics of Computation, V31, 1977.

11. R. Donati and V. Sonnad, "Coupling a Finite Element Frontal Solver Code to a Parallel Conjugate Gradient Solver", KGN 98, IBM Kingston, NY, 1986.

12. J. Winget and T. Hughes, " Solution algorithm for Nonlinear Transient Heat Conduction Analysis Employing Element-by-Element Iterative Strategies", Computer Methods in Applied Mechanics and Engineering, v52, 1985.

13. R. King and V. Sonnad, "Implementation of an Element-by-Element Algorithm for the Finite Element Method on a Loosely Coupled Array of Processors", RJ5272(54312) , IBM San Jose, 1986.
14. A. Capotondi, V. Sonnad and S. Chin, "Parallel Resolution of the Shallow Water Equations Using an Explicit Finite Difference Algorithm", KGN57, IBM Kingston, 1986.

A LOOK AT THE EVOLUTION OF MATHEMATICAL SOFTWARE FOR DENSE MATRIX PROBLEMS OVER THE PAST FIFTEEN YEARS*

J. J. Dongarra and D. C. Sorensen

Mathematics and Computer Science Division
Argonne National Laboratory
9700 South Cass Avenue
Argonne, Illinois 60439-4844

and

Center for Supercomputing Research and Development
University of Illinois at Urbana-Champaign
305 Talbot Laboratory
104 South Wright Street
Urbana, Illinois 61801-2932

1. Introduction

In this paper we look at the evolution which has taken place in the design of mathematical software for dense matrix problems. Our main emphasis is on algorithms for solving linear algebra problems where the software we develop would reside in a library on high-performance computers.

2. A Look at EISPACK and LINPACK

Both EISPACK [8,6] and LINPACK [1] are based on a *decomposition* approach to numerical linear algebra. The general idea is the following. Given a problem involving a matrix A, one factors or decomposes A into a product of simple, well structured, matrices which can be easily manipulated to solve the original problem. This divides the computational problem into two parts: first compute an appropriate decomposition; then use it to solve the problem at hand.

EISPACK is a collection of routines that compute the eigenvalues and eigenvectors of matrices and matrix systems. The algorithms in EISPACK are primarily based on Algol procedures developed in the 1960's by nineteen different authors and published in the journal *Numerische Mathematik*. J.H. Wilkinson and C. Reinsch collected a number of these procedures, together with some background material, in a volume entitled *Linear Algebra* in the *Handbook for Automatic Computation* series[9].

In early 1970 a group of researchers from Argonne National Laboratory, the University of Texas, and Stanford University proposed to the National Science Foundation a project to "explore the methodology, costs, and resources required to produce, test and disseminate high-quality mathematical software and to test, certify, disseminate, and support packages of mathematical software". The project was to take the algorithms dealing with the eigenvalue problem from the Wilkinson - Reinsch

* Work supported in part by the Applied Mathematical Sciences subprogram of the Office of Energy Research, U.S. Department of Energy, under Contracts W-31-109-Eng-38, DE-AC05-840R21400, and DE-FG02-85ER25001.

Handbook; translate them to Fortran; and produce a portable, robust, accurate, uniform, and well-documented package. At the time, little thought was given to high-performance computers; the state of the art in supercomputers was a CDC 7600. By May of 1972 the collection of Fortran routines called EISPACK was ready for public use.

A few years later there was a general feeling in the community that a unified package dealing with the solution of linear equations was needed. LINPACK is a package of routines for solving various kinds of linear systems of equations. The subroutines in the package are not the result of a translation but of a rewrite of many of the algorithms to conform to a certain style.

LINPACK uses column-oriented algorithms to preserve locality of reference. Column orientation means that the LINPACK codes always reference arrays down columns, not across rows. This approach works because Fortran stores arrays in column order. Thus, as one proceeds down a column of an array, references proceed sequentially in memory. On the other hand, as one proceeds sequentially across a row, references jumps across physical memory. The length of the jump being proportional to the length of a column. The effects of column orientation are dramatic; on systems with virtual or cache memories, the LINPACK codes will significantly outperform codes that are not column oriented.

An important factor affecting the efficiency of LINPACK is the use of the Basic Linear Algebra Subprograms (BLAS)[7]. This set of subprograms performs basic operations from linear algebra, such as computing an inner product or adding a multiple of one vector to another. In LINPACK the great majority of floating-point calculations are done within the BLAS.

One difficulty with using the BLAS, however, is that they operate only on vectors. However, the algorithms as implemented tend to do more data movement than is necessary. As a result, the performance of the routines in LINPACK suffers on high-performance computers where data movement is as costly as floating-point operations. To illustrate the point, we examine in Table 1 the peak theoretical performance of various high-performance computers available today.

Table 1
Peak Performance

Machine	Cycle time, *nsec*	Number Procs.	Peak Perf., *MFLOPS*
CONVEX C-1	100	1	20
Alliant FX/8	170	8	44
SCS-40	45	1	44
FPS 264	38	1	54
Amdahl 500	7.5	1	133
CRAY-1	12.5	1	160
CRAY X-MP-1	9.5	1	210
IBM 3090/VF-200	18.5	2	216
Amdahl 1100	7.5	1	267
NEC SX-1E	7	1	325
CDC CYBER 205	20	1	400
CRAY X-MP-2	9.5	2	420
IBM 3090/VF-400	18.5	4	432
Amdahl 1200	7.5	1	533
NEC SX-1	7	1	650
CRAY X-MP-4	9.5	4	840
Hitachi S-810/20	14	1	840
NEC SX-2	6	1	1300
CRAY-2	4.1	4	2000

By peak theoretical performance we mean that the manufactures guarantees that programs will not exceed these rates, sort of a *speed of light* for a given computer.

In practice, with LINPACK routines SGEFA and SGESL (used to solve dense systems of equations) we observe the following[3]:

Table 2
LINPACK Benchmark
Solving a 100 x 100 Matrix Problem

Machine	Peak Performance, MFLOPS	Actual Performance, MFLOPS
CONVEX C-1	20	2.9
Alliant FX/8	44	7.6 (8 proc)
SCS-40	44	7.3
FPS 264	54	5.6
Amdahl 500	133	14
CRAY-1	160	12
CRAY X-MP-1	210	24
IBM 3090/VF-200	216	12 (1 proc)
Amdahl 1100	267	16
NEC SX-1E	325	35
CDC CYBER 205	400	17
CRAY X-MP-2	420	24 (1 proc)
IBM 3090/VF-400	432	12 (1 proc)
Amdahl 1200	533	18
NEC SX-1	650	39
CRAY X-MP-4	840	24 (1 proc)
Hitachi S-810/20	840	17
NEC SX-2	1300	46
CRAY-2	2000	15 (1 proc)

At one time, a programmer had to go out of his way to code a matrix routine that would not run at nearly top efficiency on any system with an optimizing compiler. Owing to the proliferation of exotic computer architectures, this situation is no longer true. However, by using one of the new features of many modern computers namely, hierarchical memory organization one can gain high performance by some algorithmic ingenuity.

3. Restructuring Algorithms

Typically a hierarchical memory structure involves a sequence of computer memories ranging from a small, but very fast memory at the bottom to a large, but slow memory at the top. Since a particular memory in the hierarchy (call it M) is not as big as the memory at the next level (M'), only part of the information in M' will be contained in M. If a reference is made to information that is in M, then it is retrieved as usual. However, if the information is not in M, then it must be retrieved from M', with a loss of time. To avoid repeated retrieval, information is transferred from M' to M in blocks, the supposition being that if a program references an item in a particular block, the next reference is likely to be in the same block. Programs having this property are said to have *locality of reference*. Typically, there is a certain startup time associated with getting the first memory reference in a block. This startup is amortized over the block move.

If we examine the algorithm used in LINPACK and look at how the data are referenced, we see that at each step of the factorization process there are vector operations that modify a full submatrix of data. This update causes a block of data to be read, updated, and written back to central memory. The number of floating point operations is $\frac{2}{3}n^3$ and the number of data references, both loads and stores, is $\frac{2}{3}n^3$. Thus, for every *add/multiply* pair we must perform a load and store of the elements, unfortunately obtaining no reuse of data. Even though the operations are fully vectorized, there is a significant bottleneck in data movement, resulting in poor performance (see Table 2). On vector computers this translates into 2 vector operations and 3 vector-memory references. Usually limiting the performance to well below peak rates. To achieve high-performance rates, this *operation to memory reference rate* must be higher.

It is possible to express the algorithms in terms of matrix-vector operations. These operations have the benefit that they can reuse data and achieve a higher rate of execution than the vector counterpart. In fact, the number of floating-point operations remains the same; only the data reference pattern is changed. This change results in a operation to memory reference rate on vector computers of effectively 2 vector operations and 1 vector-memory reference.

When the algorithm is recast to minimize memory traffic, the results can be impressive. In [2] the algorithm for solving dense systems of linear equations was recast in terms of a matrix-vector multiply, which has the desired effect (see Table 3).

Table 3.
Comparison with Matrix-Vector Operations
Solving a 100 x 100 Matrix Problem

Machine	Performance Before MFLOPS	Performance After MFLOPS
CONVEX C-1	2.9	5
Alliant FX/8	7.6 (8 procs)	10
SCS-40	7.3	13
FPS 264	5.6	24
CRAY-1	12	38
CRAY X-MP-1	24	57
IBM 3090/VF-200	12	24(1 proc)
Amdahl 1100	16	48
NEC SX-1E	35	71
CDC CYBER 205	17	24
CRAY X-MP-2	24 (1 proc)	48 (2 procs)
IBM 3090/VF-400	12	24(1 proc)
Amdahl 1200	18	52
NEC SX-1	39	74
CRAY X-MP-4	24 (1 proc)	74 (4 procs)
Hitachi S-810/20	17	48
NEC SX-2	46	100
CRAY-2	14	28 (1 proc)

The same effect can be reallized for many of the algorithms dealing with other matrix problems, such as eigenvalue problems[4].

Table 4 shows the comparison between the original EISPACK formulation and the restructured matrix-vector algorithms.

Table 4
Speedup of Matrix-Vector Versions over the EISPACK Routines

Routine	order 50	order 100	Machine
ELMHES	1.5	2.2	CRAY 1
ORTHES (1)	2.0	1.9	CRAY 1
ORTHES (2)	2.5	2.5	CRAY 1
ELMBAK	2.2	2.6	CRAY 1
ORTBAK (1)	2.8	2.5	CRAY 1
ORTBAK (2)	3.6	3.3	CRAY 1
TRED1	1.5	1.5	CRAY X-MP-1
TRBAK1	4.2	3.7	CRAY X-MP-1
TRED2	1.6	1.6	CRAY X-MP-1
SVD no/v	1.7	2.0	Hitachi S-810/20
SVD w/v	1.6	1.7	Hitachi S-810/20
REDUC	1.8	2.2	Fujitsu VP-200
REBAK	4.4	5.8	Fujitsu VP-200

All the data represent Fortran implementations; no attempt was make to explicitly exploit the vector hardware through assembly language.

If we look at the current generation of high-performance computers on the market today, we see even more emphasis on memory hierarchies. Machines such as the CRAY-2, Alliant FX/8, and IBM 3090/VF all have an additional level of memory between the main memory and the vector registers of the processor. This memory, referred to as *cache* or *local memory* is relatively small (on the order of 16K words) and may not be under the control of the programmer.

Nevertheless, the issues are the same: to come close to gaining peak performance, one must optimize the use of this level of memory (i.e. retain information as long as possible before the next access to main memory) obtaining as much reuse as possible. In the case of matrix factorization this means that instead of *matrix-vector* operations, one must perform *matrix-matrix* operations[5]. Here we can achieve a high operation to memory reference rate. In the case of matrix-matrix operations we get a *surface to volume* effect, $O(mn^2)$ operations to $O(mn)$ memory references. An effort is currently under way to recast the algorithms in terms of these Level 3 BLAS, matrix-matrix operations. This involves modifying the algorithm to perform more than one step of the decomposition process at a given loop iteration.

4. Conclusions

In an attempt to easily transport algorithms to a wide variety of architectures and to achieve high performance, we are isolating the computationally intense parts in high-level modules. When the architecture changes, we deal with the modules separately, rewriting them in terms of machine-specific operations; however, the basic algorithm remains the same. By doing so we can achieve the goal of a high operation to memory reference ratio.

1. J. Dongarra, J. Bunch, C. Moler, and G. Stewart, *LINPACK Users' Guide,* SIAM Pub, Philadelpha (1976).

2. J. Dongarra and Stanley C. Eisenstat, "Squeezing the Most out of an Algorithm in Cray Fortran," *ACM Trans. Math. Software* **10, 3**, pp. 221-230 (1984).

3. J.J. Dongarra, "Performance of Various Computers Using Standard Linear Equations Software in a Fortran Environment," Argonne National Laboratory MCS-TM-23 (October 1986).

4. J.J. Dongarra, L. Kaufman, and S. Hammarling, "Squeezing the Most Out of High Performance Computers for Finding the Eigenvalues," *Linear Algebra and Its Applications* **77**, pp. 113-136 (1986).

5. J.J. Dongarra and D.C. Sorensen, "Linear Algebra on High-Performance Computers," pp. 3-32 in *Proceedings Parallel Computing 85*, ed. U. Schendel, North Holland (1986).

6. B.S. Garbow, J.M. Boyle, J.J. Dongarra, and C.B. Moler, *Matrix Eigensystem Routines - EISPACK Guide Extension*, 1977.

7. C. Lawson, R. Hanson, D. Kincaid, and and F. Krogh, "Basic Linear Algebra Subprograms for Fortran Usage," *ACM Transactions on Mathematical Software*, pp. 308-323 (1979).

8. B.T. Smith, J.M. Boyle, J.J. Dongarra, B.S. Garbow, Y. Ikebe, V. Klema, and C. Moler, *Matrix Eigensystem Routines - EISPACK Guide, Second Edition,*, 1976.

9. J. Wilkinson and C. Reinsch, *Handbook for Automatic Computation: Volume II - Linear Algebra,* Springer-Verlag, New York

A Graphical Approach to Load Balancing and Sparse Matrix Vector Multiplication on the Hypercube

Geoffrey C. Fox †
California Institute of Technology
Pasadena, CA 91125
December 5, 1986

Presented at Minnesota Institute for Mathematics and Its Applications Workshop November 6, 1986

Abstract

We consider the implementation on a hypercube concurrent computer, of matrix vector multiplication $y=Ax$ where A is a large sparse matrix. A good decomposition is crucial for the case when each column of A has on the average fewer non zero elements than there are nodes in the hypercube. We review simulated annealing and neural network methods for generating a hypercube decomposition for this problem. We introduce a new graphical method, orthogonal recursive bisection, which can be applied to general problems and is successful on this test case. The performance of the concurrent computer depends strongly on any correlations in the placement of non zero elements of A.

I: Introduction

We consider the solution of a very simple problem - namely matrix A times vector $\underline{x}$ multiplication, i.e.

$$\underline{y} = A\underline{x} \tag{1}$$

on a hypercube concurrent computer. We will see that this seemingly trivial problem is intellectually challenging and exhibits the essential features of a wide class of network and time synchronized simulations on the hypercube. The techniques developed apply to any distributed memory coarse grain sized computer but here we will specialize to the hypercube topology in the implementations and detailed discussion.

We wish to consider the case when A is very sparse. An example we will often use is the sparse matrix coming from the simplest five point discretization of the Laplacian operator ∇^2 in two or three dimensions. Then for a 1000×1000 regular mesh, $\underline{x}$ and $\underline{y}$ have 10^6 elements. A has 10^6 rows and columns but at most 4 nonzero elements in each row or column. In this example A is symmetric although the techniques that we discuss will not depend on this special feature. We will let M be the length of the vectors $\underline{x}$ and $\underline{y}$ and suppose that each row or column of A has an average of L nonzero elements; i.e. the $M\times M$ matrix A has a total of ML nonzero elements. Except for pathological cases the performance of a hypercube on (1) depends on the average number of nonzero elements and not, say, or the maximum number in any one row or column. On the other hand we will see that the actual positioning of the

† Work supported in part by DOE grant DE-FG03-85ER25009, Parsons and System Development Foundations, and the office of the program manager of the Joint Tactical Fusion Office.

nonzero elements of A, i.e. correlations between these positions, can be quite crucial in determining performance.

There are several approaches to the implementation of (1) on the hypercube; let us consider one generally good and simple technique. We decompose the vectors $\underline{x}$ and $\underline{y}$ so that each processor holds M/N elements of each vector. Here the hypercube of dimension d_H has

$$N=2^{d_H} \tag{2}$$

processors or nodes. There are, of course, many $({}_NC_{M/N})$ ways of decomposing M objects over N nodes and in fact the key problem discussed in this paper will be finding which of the many decompositions to choose. As there are so many possibilities, an exhaustive search is impossible; we need to find ways of finding a good, but not necessarily the best, decomposition from a search that only samples over a few possible decompositions. The problem is NP complete in the language of computer science.

Once one has chosen a decomposition, the concurrent algorithm is simple. Each node stores both the nonzero values (locations) of A_{ij} and A_{ji} for all vector locations (values) $j(x_j)$ held in this node. The algorithm divides into two stages.

I: Transmit from node α the value of x_j to node β where the index j is associated with node α, index i with node β and A_{ij} is nonzero (3-I)

II: Calculate within each node the summation (1). (3-II)

II involves no communication and is completely load balanced. I is the communication which needs to be minimized. Note that this communication overhead is a distinctive consequence of the distributed memory architecture and if we are successful in reducing it, we are showing the relevance of this architecture to this class of problems.

We would also note that the issues behind the steps (3-I), (3-II) are much more general and important than the simple problem (1). Consider for instance a particle dynamics problem where we wish to evolve in time a set of M particles. At any one time, particle i is at position $\underline{r}_i$. Define the matrix A_{ij} to be non zero if and only if particles i and j interact with each other; typically one would have

$$A_{ij} \neq 0 \quad \text{if and only if } |\underline{r}_i - \underline{r}_j| < R \tag{4}$$

where R is the range of the force between the particles. Then the decomposition of this problem is very similar to that of (1); we wish to distribute particles i over the hypercube in such a way that we minimize the number of nonzero matrix elements A_{ij} that cross processor boundaries. In the problem (1), A_{ij} is fixed whereas in the particle dynamics case, A_{ij} changes as the system evolves. The static decomposition and dynamic load balancing problems are essentially identical. In fact, any time stepped simulation can be reduced to the study of a, in general, sparse matrix A_{ij} specifying the interconnection between elements i and j in the simulation. A nontrivial complication for a general simulation is the presence of elements i of varying nature i.e. different computational complexity; this complicates but does not seem to change the basic issues. We will review the general case in detail elsewhere (Ref. 1) but will for the rest of the paper concentrate on the specific problem (1).

In section II, we define the algorithm (3) more precisely and calculate the performance of the hypercube for a given decomposition. In sections III and IV we cover three decomposition techniques; section III discusses simulated annealing[2,3] and neural nets[4] and IV a new graphical method. In the final section, we discuss tests of the new method and suggest further research issues.

II Calculation and Communication Times on the Hypercube

In this section, we assume that we have created a map on the hypercube of

$$\text{vector element } i \rightarrow \text{processor } P_i \tag{5}$$

and analyze the behavior of the hypercube on the problem (1) for the decomposition (5). Let take use as usual the notation

t_{comm}: communication time between adjacent nodes on the hypercube (6a)

for a 64 bit quantity

t_{flop}: typical time to add or multiply two 64 bit(double precision) numbers. (6b)

We will assume that equation (1) is to be calculated in double precision; the analysis is essentially identical if this is not true. One often uses the ratio given by:

$$\tau = t_{comm}/t_{flop} \tag{6c}$$

Let us now consider first the calculation stage (3-II). Equation (1) involves a total of $M(2L-1)$ floating point operations; so the sequential execution time is

$$T_{seq} = M(2L-1)t_{flop} \tag{7-Seq}$$

The stage (3-II) is under assumptions explained below load balanced and so for the hypercube implementation

$$T_{conc}^{II} = \left[M(2L-1)/N\right] t_{flop} \tag{7-II}$$

To be precise, stage (3-II) is only exactly load balanced for an assignment of equal number of M/N elements per processor when L, the number of non zero elements per row in A_{ij} is independent of i. In many cases, L is in fact essentially constant and so (7-II) is accurate. In general, one needs to allow L to vary and the load balanced decomposition to allow different number of elements in each node. This corresponds to the case mentioned in the introduction where the computational complexity C_i depends on the element i. The techniques described in sections III and IV can be extended to variable C_i and unless C_i takes extreme values, we do not believe it affects the essential issues. Thus, in this paper, we will only consider the case when C_i is a constant $(2L-1)$ independent of i.

The communication cost for stage (3-I) depends greatly on circumstances. For the example discussed in the introduction when A comes from the discretization of the Laplacian in two dimensions, one can use a simple domain decomposition of the type shown in figure 1. Now i labels the points in a simple two dimensional grid. Only for elements i on the edge of the subdomain held in a particular processor is any communication necessary. Thus one finds a communication time T_{conc}^{I} that is proportional to $\tau\sqrt{N/M}$ times the calculational time T_{conc}^{II} given in (7-II). In the interesting limit $M \gg N$, calculation dominates communication for machines like the Caltech/JPL hypercube for which the ratio $\tau \sim 1.5$ is not large compared to unity. For such an A it is possible to group elements so that only little communication is necessary . Now let us consider what would happen if we had not chosen such a sensible decomposition or if the non zero elements in the matrix A did not cluster in a way that allowed this type of decomposition.

For a typical non-optimized decomposition one must expect that in the sum $\sum_{\substack{j \\ L\ \text{values}}} A_{ij} x_j$, the L interesting x_j will not be stored in P_i; the processor associated with element i. In fact, each x_j only has a chance $1/N$ of being found in P_i. On the hypercube the average communication distance is $\frac{1}{2}\log_2 N$ and so a typical non optimized decomposition will demand a total communication time T_{comm} of $\frac{ML}{2}\log_2 N$. In general, one would expect this communication to be balanced and so the worst case scenario is

$$T^I_{conc} = \frac{T_{comm}}{N} = \frac{M}{N}\frac{L}{2}\log_2 N t_{comm} \tag{7-I}$$

The speedup on the hypercube, for this non optimized decomposition, is:

$$\begin{aligned} S &= \frac{T_{seq}}{T^I_{conc} + T^{II}_{conc}} \\ &\sim \frac{N}{\log_2 N}\,\frac{2(2L-1)}{L\tau} \end{aligned} \tag{8}$$

for large N where T^I dominates T^{II}. The performance (8) still represent substantial speed up but it is poor compared to the form

$$S \sim N/\left(1 + \frac{4\tau}{7}\sqrt{N/M}\right) \tag{9}$$

for the decomposition given in figure 1. In this paper, we will be discussing ways of improving (8). We will devise general methods that generate the result (9) (or nearly it) for the simple case of figure 1 but which generalize to more complicated matrices A.

We will first discuss the communication in more detail. As a first simplification, we note that we will ignore any imbalance in communication and only discuss the total communication time T_{comm} summed over all transmitted vectors x_j. This approximation needs to be examined in more detail but we believe that it does not miss any essential issues.

x_j must be sent from node P_j to all nodes P_i for which A_{ij} is nonzero; this is a list of up to L distinct nodes to which x_j must be sent. This is a variant of the standard traveling salesman problem which can be solved more efficiently than the naive estimate (7-I). In appendix A, we detail the algorithm we used in our numerical tests. This does not give the best routing but it is quite good and fast to generate. As we will see in sections IV, this change from L separate trips to one coordinated trip reduces the estimate (7-I) by approximately a factor of 2 in the case L=4. One simple limit is the case when L is $\gtrsim N$ and here the maximum trip distance for x_j to visit its L associated nodes is $N-1$. This is the well known long range force algorithm when all information is sent to all nodes. In this case, the speed up becomes

$$S \sim N/\left(1 + \frac{\tau(N-1)}{2L-1}\right) \tag{10}$$

which gets arbitrarily close to N as L increases and becomes large compared to N. The simple long range force problem corresponds to a full matrix A and $L=M$.

We should also note that the concurrent algorithm described here and in section I is not the only one possible. Another reasonable strategy is to form in each node the partial sums $\sum_{\substack{j \text{ in} \\ \text{same node}}} A_{ij} x_j$ and route these to the target node P_i. As these partial sums, circulate through the hypercube, those corresponding to the same element i are combined as they transit through the same intermediate node. Good algorithms for this have been presented for full matrix vector multiplication and other problems in Ref. 5 and studied by Furmanski[6].

III Simulated Annealing

Simulated annealing is a powerful method for optimizing functions defined over complex systems which has been introduced by Kirkpatrick and colleagues[7]. We have shown in a series of papers[2,3,8] how to apply it to load balancing on the hypercube for problems of the type given in equation (1). This method is derived by mapping the complex system represented by (1) into a physical system.

We consider each vector element i as a particle moving in space with $N = 2^{d_H}$ distinct positions. These particles interact with two types of forces[2,3,8]. These forces are represented by a energy or objective function E which, therefore, has two terms. The first term corresponds to a short range repulsive force between the particles and takes the form

$$E_1 = \sum_P \left(\sum_{i:P_i=P} C_i \right)^2 \tag{11}$$

where P runs over the N positions in the space. $\sum_{i:P_i=P} C_i$ is the total calculational complexity for the concurrent algorithm in processor P. E_1 can be minimized in our case by always ensuring that each position (processor) contains an equal number of particles and this constraint will either be exactly or approximately embodied in our methods.

The most interesting term in the energy function corresponds to an attractive force between any two particles i, j that correspond to a non zero entry A_{ij} in the matrix A. In fact

$$E_2 = T_{comm} / t_{comm} = \sum_i V_i \tag{12}$$

where T_{comm} is the total communication time introduced in section II. V_i is the total route length involved in transmitting particle i from P_i to all nodes P_j for which A_{ji} is nonzero. The exact form of V_i used here is given in appendix A. If the simple non optimal routing of (7-I) is used, V_i just becomes a sum of two particle potentials V_{ij} where,

$$V_i = \sum_{j:A_{ji} \neq 0} [V_{ij} = \text{distance from } P_i \text{ to } P_j \text{ in hypercube}] \tag{13}$$

the individual V_{ij} are just linear (in hypercubic space!) potentials. V_{ij} is clearly attractive and it is minimized when i and j are at the same position (i.e. in the same processor). Although the form V_i in (13) is not precise, it does give one a reliable intuition as to the nature of the interactions between the particles. In this physics analogy, we need to find the ground state of the system of particles. Translating this back to the original complex system (1), this ground state then corresponds to the required optimal decomposition of elements i onto the hypercube.

There are some well established techniques for finding the equilibrium state of physical

system which may be applied to the physics analogy; this is the method of simulated annealing. We will describe the Metropolis method which is one of the most powerful and was used in Ref. 3 for problems similar to those discussed here. We assume E_1 is minimized by keeping equal number of elements in each node.
Now define

$$f = E_2/MT = 1/T\left[\sum_i V_i\right]/M \tag{14}$$

where T is a temperature. The attractive potential energy E_2 and hence f depend on the particular configuration (map) of particles (elements) i into positions (processors) P_i. Consider any one such map.

$$M: i \rightarrow P_i \text{ with energy } E_2 \tag{15a}$$

Generate from this a new map

$$M': i \rightarrow P_i' \text{with energy } E_2' \tag{15b}$$

where M' is gotten by interchanging any two particles i and j i.e.

$$P_i' = P_j \tag{16}$$

$$P_j' = P_i$$

$$\text{Let } \delta f = f' - f = (E_2' - E_2)/MT \tag{17}$$

Note that δf is easily calculable as most of the terms V_i in E_2' and E_2 are identical. At most $2L+2$ terms are different between E_2 and E_2' corresponding to V_i, V_j and any V_k such that i or j are involved in its calculation i.e. such that particle k interacts with i and j. The Metropolis algorithm is defined by:

i) if $\delta f \leq 0$, replace configuration M by M' (18)

ii) if $\delta f > 0$, replace M by M' with probability $\exp[-\delta f]$.

when the temperature T is large, δf is small and essentially all changes are accepted; as the system is cooled and T is reduced, one tends to accept only changes that reduce the energy and so improve the decomposition. For the parameter $L=4$, we found that $T=1$ allowed one to sample many configurations and find a good starting point. Then one could lower T to 0.25 or 0.5 and converge to an approximate ground state. We will use some the results from this method in section V.

We have recently found that the neural network method based on the work of Hopfield and Tank[9] is often superior to simulated annealing. This new method is quite similar to annealing and can be viewed as a deterministic "equations of motion" method of finding the ground state of the analog physical system introduced above.

We have only specified the potential energy above and the physical system can be completed by many different kinetic terms. Further such systems are typically conservative and evolving their equations of motion would not lead to the ground state but rather a constant energy orbit. We use biological intuition to suggest an appropriate set of equations which have an effective frictional term to cause the system to evolve to minimum of the potential energy $E = E_1 + const.\ E_2$. Different choices of equations will lead to different speeds of convergence and different probabilities of reaching the true rather than a local minima.

We first introduce "neural variables" associating $d_H = \log_2 N$ of them with each point i. Write the processor associated with i as:

$$P_i = \sum_{c=0}^{d_H-1} 2^c [1+S_c(i)]/2 \tag{19}$$

where the neurons (spins) $S_c(i)$ take the values ± 1. Clearly, $S_c(i) = +1$ implies that the associated processor P_i is "up" in the cth direction in the hypercube. The replacement of P_i by the set $\{S_c(i)\}$ is a useful step even the simulated annealing approach. We will find iteratively the values $S_o(i)$ (for all i),$S_1(i) \cdots S_{d_H-1}(i)$. This leads to $\log_2 N$ stages, at each of which we only have two choices; this is faster than Eq. (16) with one stage consisting of N choices for P_i.

Expressing E in terms of $S_c(i)$ we can use the mean field approximation to calculate the statistical physics particular function $Z = \sum \exp(-E/T)$ associated with the spin system $\{S_c(i)\}$. This leads to a Hamiltonian

$$H = \frac{E}{T} + \sum_{c,i} \left\{ u_c(i) S_c(i) - \ln \cosh u_c(i) \right\} \tag{20}$$

with mean field equations

$$\frac{\partial H}{\partial u_c(i)} = S_c(i) - \tanh u_c(i) = 0 \tag{21}$$

$$\frac{\partial H}{\partial S_c(i)} = u_c + \frac{\partial E}{T \partial S_c(i)} = 0$$

to find the stationary points in H. $S_c(i)$ and $u_c(i)$ can be considered conjugate positions and momenta in a Hamiltonian formalism.

As discussed in Refs. [9] and [4], there are various methods of solving (21) which have different performance or different problems. We introduced in Ref. [4], the *bold* network which was both fast and robust for hypercube decompositions. A key advantage of neural networks is that is is rather straightforward to estimate their speed of convergence. The total time taken to solve (21) is

$$T^{\text{neural net}} = c(\eta) M^{1+1/d_s} \tag{22}$$

where the value of $c(\eta)$ depends on the required goodness of convergence; this is the time taken to reach a value of E that is a factor $(1+\eta)$ times its minimum value. A value of $\eta \sim 0.2$ would be a typical goal. In equation (22), d_s is the dimension of the underlying system. A general definition of this is given in Ref. [2]. The factor M^{1/d_s} represents the maximum distance between points in the graph A_{ij}. As discussed in Ref. [4], there is some reason to believe that one can improve (22) to:

$$T^{\text{neural net}} = c(\eta) M \ln M \tag{23}$$

Simulating annealing can be in principle always get a solution that is arbitrary near the true maximum. However this is at a cost - in terms of time spent annealing - that is arbitrarily great. Neural nets seem to give generally good results in a time, given in equations (22) and (23), that is reasonable and more importantly predictable.

IV ORB: Orthogonal Recursive Bisection

IVA The Graphical Distance $d(i,j)$

One problem with the methods of section III is that they are time consuming especially when M, the number of elements, is large; the statistical nature of the procedure leads to slow convergence to equilibrium for simulated annealing. Neural networks are faster but the work needed to generate a decomposition grows with the size of the systems to a power greater than one. Here we present a different technique that will not produce such a good decomposition but one which in some cases, is quite good. Further the method is deterministic and requires modest calculation which is linear in the system size.

We consider A_{ij} as defining a graph with nodal points labeled by i and non zero entries in A_{ij} specifying connected nodal points. The connections are bidirectional if A is symmetric but in general are directed. We will use a physics notation and let $|i>$ denote an arbitrary nodal point in the graph. A is an operator that maps links $|i>$ to a set $\{|j>\}$. Define the distance $d(i,j)$ between $|i>$ and a general point $|j>$ by

$$|j> \underset{\min}{=} (A+A^T)^{d(i,j)}|i> \qquad (24)$$

where = in the above equation should be read as "|j> is included in the set of points gotten by acting $A+A^T$ (A^T is transpose of A) on $|i>$." (24) indicates that one chooses $d(i,j)$ to be the smallest integer power of $A+A^T$ for which $|j>$ appears in list of generated points.

The above definition of $d(i,j)$ can be easily interpreted for the case when A corresponds to the discretized Laplacian on a two dimensional mesh. As illustrated in figure 2, $d(i,j)$ corresponds to the two dimensional "stair-case" rather than the conventional direct distance.

We wish to find a way of decomposing the graph (set of elements of i) which gives reasonable results for the two dimensional space (i.e. results competitive with figure 1) but only uses information contained in A and not the "secret" information as to the existence of an underlying space. Roughly this means we only used the distance $d(i,j)$ defined in (24) rather than the complete structure of the underlying metric space. We will use a recursive method that has been applied to metric based decompositions before but not as far as we know to general graphs.

The basic idea is to divide the graph into N subgraphs such that nearby points (i.e. points $|i>$, $|j>$ for which $d(i,j)$ is small) are in the same subgraph. This procedure does not precisely implement the goals defined in section II, the minimization of T_{comm}, but is is qualitatively similar. We expect that if we can find decompositions that minimize $d(i,j)$ when $|i>$, $|j>$ are in same processor, then these are also be near the best in the sense of section II.

We will first describe the *ring algorithm* which is not optimal but illustrates some of the basic ideas.

IVB The Ring Algorithm

Given a graph as defined above, let $|i_B>$ be a random initial nodal point. Associate with each point $|j>$ the distance

$$d_B(j) = d(i_B,j) \qquad (25a)$$

Let j_{ext} be the label of one of the nodal points that maximizes $d_B(j)$. Figure 3 illustrates this for the graph corresponding to a 16×16 two dimensional mesh. j_{ext} is one of the four corners of the domain. The corner chosen will depend on particular point $|i_B>$ picked. Now form the ring distance

$$d_{ring}(j) = d(j_{ext}, j) \quad (25b)$$

This is the staircase distance measured from the corner $|j_{ext}>$. We now sort the set $\{|j>\}$ on the basis of the function $d_{ring}(j)$ and divide $\{|j>\}$ into N equal subsets based on this ordering. This divides the graph up into rings centered on $|j_{ext}>$. This is also illustrated in Figure 3. We obtain a hypercube decomposition by mapping the hypercube into a one dimensional system as again shown in figure 3.

The *ring algorithm* is appropriate for a concurrent computer with a one dimensional mesh interconnect but nonoptimal for a hypercube. It produces long skinny subdomains which involve more communication than the natural square (hexagonal .. circular) subdomains which have lower edge/area ratios and hence minimal communication.

IVC The Bisection Algorithm

We can modify the *ring algorithm* in a simple fashion which leads to much better results. The bisection algorithm is recursive and at each stage is dealing with a subgraph of the original graph. When considering a subgraph, we define the restricted distance $d_R(i,j)$ using a restricted operator A defined so that the "staircase" from $|i>$ to $|j>$ only uses intermediate nodal points that lie within in the subgraph. The recursive method used by ORB is essentially that mentioned at the end of section III as the best approach to the neural network and simulated annealing approach.

Consider a given subgraph. Then execute the algorithm IVB to define $d_R(j_{ext},j)$. Now rather than dividing the system into N parts, just divide it into two. This is illustrated in Figures 4(a) and 5(a). In 4(a) we show the first stage in the bisection algorithm. Now that the graph is divided into two subgraphs, one recursively applies the same algorithm to each subgraph. The recursion uses $\log_2 N$ stages with a total of 2^k subgraphs being generated after the kth stage is completed. In figure 5(a), we show the final result of this process for the same problem - a 16×16 mesh - used for the ring algorithm in figure 3. We see that we have formed subdomains which are better shaped with lower communication. In fact, as record in table 1, the simple rectangular decomposition on an 8 node hypercube leads for $M = 256$ to

$$T_{comm} = 0.5Mt_{comm} \quad (26a)$$

where as the *bisection algorithm* leads to

$$T_{comm} = 0.57Mt_{comm} \quad (26b)$$

and the *ring algorithm* to

$$T_{comm} = 0.66Mt_{comm} \quad (26c)$$

There is actually an in general important technical point that we have ignored in the above discussion. Thus, the distance $d(i,j)$ or even importantly $d_R(i,j)$ is only defined for points $|i>$ and $|j>$ that are connected by the full (or restricted) operator A. In general, given a subgraph, one divides it into connected sectors of points internally linked by A. In the current algorithm, we sort these sectors in order of decreasing size. Each sector is ordered by $d_R(j_{ext},j)$ (*ORB*) or $d(j_{ext},j)$ (*ring*) and then we assign alternate sectors to the beginning and end of a list which is then either bisected (ORB) or divided into N parts (*ring*)

IVD Orthogonal Recursive Bisection

It is clear from figure 5(a) that the bisection algorithm has one obvious deficiency. Namely once we have divided a graph into two by a line such as that in figure 4(a), we would expect that it would be best to choose the division at the next level of recursion to be in some sense perpendicular to the first cut. We have substantial freedom in defining a cut even within the algorithm of IVB. Thus different choices of the initial point i_B can lead to different j_{ext} and hence to cuts in very different directions. The ambiguity is irrelevant in the *ring algorithm* or in the first stage of the *bisection algorithm* but it can make important differences at later stages in the recursion.

We will in fact change the basic bisection algorithm as follows. This only applies to stages later than the first. Identify all boundary points. These are all points $|j_{boundary}>$ in the subgraph such that $(A + A^T)|j_{boundary}>$ contains a point in the other subgraph formed in the bisection that gave the given subgraph. These are marked by circles in figure 4(b). Choose any one boundary point $|j_R>$. Let $|j_A>$ be a boundary point such that $d_R(j_R, j_A)$ is maximized. Now order all points with respect to $|j_A>$. Let the boundary point $|j_B>$ maximize $d_R(j_A, j_B)$. This seemingly complex procedure is intuitively simple. We wish to form a line of boundary points and find a beginning $|j_A>$ and end $|j_B>$ point on the boundary. In its simplest form, the new algorithm now orders all points $|j>$ in the graph by the difference in distances

$$\delta d = d_R(j_A, j) - d_R(j_B, j) \tag{27}$$

This intuitively will divide subgraph into two with a cut that divides the boundary in two and, initially at least, the new cut is orthogonal to the boundary which represents the previous cut. Thus we term the *orthogonal recursive bisection* or ORB algorithm.

Figure 5(b) shows this clearly where the algorithm of this section has moved the third cut from its position in figure 5(a). However note that it only is orthogonal to the second cut near the boundary. It tails away so that for instance processor 1 holds a strangely shaped domain. This may not be a serious problem but we can extend algorithm the in the following interesting fashion. Replace δd in Eq. (27) by

$$\delta d = \sum_{k=1}^{P_{cut}} \left[d_R(j_{A_k}, j) - d_k(j_{B_k}, j) \right] \tag{28}$$

Here $A_1 = A$ and $B_1 = B$ are the original extrema used in (27). The idea here is to keep the new cut "straight' and avoid it tailing away by replacing the single points A and B by clusters which should provide stronger distance constraints. We define A_{l+1}, B_{l+1} used in Eq. (28) recursively as the points that minimize/maximize respectively the expression

$$\delta d_l = \sum_{k=1}^{l} \left[d_R(j_{A_k}, j) - d_R(j_{B_k}, j) \right] \tag{29}$$

so that (27) and (28) are given respectively by the values $l=1$ and $l=P_{cut}$ in equation (29).

The single parameter P_{cut} defines the bisection stage of ORB. We let $P_{cut}=0$ denote the non orthogonal algorithm of section IVC and $P_{cut}=1$ equation (27).

Conventionally we use $P_{cut}=-1$ to denote an algorithm where δd is calculated from (29) where l is made as large as the subgraph allows. This extreme algorithm can produce strange results as shown in figure 5(d). Use of $P_{cut}=-1$ means that all trace of the boundary is lost and the new cut no longer bisects it.

We should also note that as explained in section IVC, this algorithm is not actually applied directly to the full subgraph but separately to each connected sector within it.

IVE Alignment

There is on important aspect of the ORB algorithm which remains to be discussed. We have divided the graph into N subgraphs but not shown how to assign each subgraph to a particular processor of the hypercube. Remember the algorithms in IVC and IVB have $\log_2 N$ stages. At the kth stage we align the 2^k subgraphs of this stage. For instance in the examples of figure with $N_{proc}=8$, at the first cut we define two subgraphs; one will be shared by processors 0, 1, 2, 3 and the other by 4, 5, 6, 7. The next cut gives four subgraphs to be assigned to the pairs 0,1 2,3 4,5 and 6,7. After the final cut, we get a unique association between subgraph and processor.

The alignment algorithm is controlled by a single parameter P_{align}. We have just divided a subgraph into two; we wish to assign one half to be "lower" (0-bit) and one to be "upper" (1-bit) in our recursive generation of a hypercube. If this is the first subgraph to be generated at this bisection stage, we make an arbitrary assignment of upper and lower. Otherwise we compare this subgraph with previous subgraphs that already have been divided. Let us label the two halves of the current subgraph [A] and [B]. Let us define a distance $D(X,Y)$ between two subgraph or subgraphs X and Y. Let [0] be the subgraph consisting of all points assigned "lower" at this bisection stage and belonging to a neighboring node in the hypercube topology. [1] is the corresponding "upper" subgraph.

Let

$$
\begin{aligned}
D_0 &= D([B], [0]) \qquad (30)\\
D_1 &= D([B], [1])\\
D_2 &= D([A], [0])\\
D_3 &= D([A], [1])
\end{aligned}
$$

We want to label [A] and [B] upper or lower, so as to align them best with the previous assignments. We label [B] lower, [A] upper if D_0 or D_3 is the minimum value of the four distances. We reverse the assignment is D_1 or D_2 is the minimum. This is not a unique algorithm and as we will comment in section V, it is one of the weaker parts of ORB.

The value of P_{align} defines the method of calculation of D. The case $P_{align}=1$ is time consuming but simple to explain. $D(X,Y)$ is just the sum over elements $i \in X$, $j \in Y$ of the point distances $d_R(i,j)$. These distances are calculated over the restricted domain gotten from the union of [A] [B], [0] and [1]. $P_{align}=0$ is a much faster calculation and almost as reliable; it is illustrated in figure 4(c). $D([A(B)],Y)$ is just calculated as the sum of all points $j \in Y$ of the distances $d_R(A(B),j)$ from the single extreme point $A(B)$ in $[A]([B])$. A and B are the extrema found in the bisection stage from minimizing or maximizing (29).

The alignment algorithm is not perfect - for instance in figure 5(a) one would "expect" the domains assigned to processors 6 and 7 to be reversed. The algorithm also has some niceties in its implementation to cope with the case when the subgraphs contain disconnected sectors. We will suggest in section V that it may be better to use simulated annealing or neural networks to improve the alignment.

We should comment on the timing of ORB. In the normal case of $P_{align} = 0$ and $P_{cut}=0$ or a small positive integer, we have the excellent result:

$$T^{ORB} = const.\ M \log_2 N \tag{31}$$

The choices $P_{align} = 1$ or $P_{cut} = -1$ involve times proportional to the square of the number of points and as shown in section V, this large time increase is not warranted by any improvement in the balancing. Also in these cases, the neural network method will typically run faster than ORB. For the simple case (31), the timing of ORB is formally lower than that of the algorithms in section III.

V Results and Discussion

We have run a limited set of tests of the new method and our results are contained in tables 1, 2 and figures 3, 5, 6, 7, 8. We have chosen three classes of matrix A:

A_i) The matrices A corresponding to a Laplacian discretized on a square two dimensional mesh with either a 16×16 or 64×64 grid.

A_{ii}) The matrix A with 544 rows studied in Ref. 3. This corresponds to the finite element solution of an irregular two dimensional region with a concentration of mesh points in a small subregion.

A_{iii}) Matrices with a random distribution of elements and either 256 or 4096 rows.

In each case the value of L, the number of nonzero elements in each row takes the value 4.

We can compare four decompositions:

D_i) The best (or nearly best) decompositions that are known for cases A_i, A_{ii}). For A_{ii}), this comes from simulated annealing.

D_{ii}) The Ring decomposition of section IVC.

D_{iii}) The ORB decomposition with various parameters P_{cut} and P_{align}.

D_{iv}) A random decomposition where we arbitrarily divide the M rows into N equal sets.

We have not explored simulated annealing and neural networks in detail here because we know that then will give a good decompositions for this class of problem[2,3,4,8].

In Table 1, we study mesh problems A_i). We find that for small N, ORB is almost perfect and even for N=64 the ORB decompositions only lead to decompositions that have about 50% higher communication than the optimal case. The method works as well on large 4096 rows as small matrices. As expected, ORB always performs significantly better than the *ring algorithm*; both are far superior to the random decomposition. The preferred parameter values appear to be P_{cut}=16 and P_{align}=0 or 1, but all solutions in table 1 are quite good. Table 1(c) applies ORB to the case A_{ii}) and is very encouraging. ORB appears to work as well for irregular as regular meshes. It gives a solution that is quite competitive with that obtained by simulated annealing in Figure 6 taken from Ref. 3. Figure 7 and 8 show the ORB solutions with P_{cut}= 16 and P_{align}=0 and 1 respectively. These last two have the same division into 16 sectors but differ in the alignment stage i.e. in the assignment of sectors to hypercube nodes. We see from table 1(c) that the more sophisticated P_{align}=1 algorithm gives the better result.

In Table 2, we consider random matrices. Even here, ORB can improve over a purely random assignment although only by about 20%. In this table, the parameters $P_{cut}=-1$, $P_{align}=1$ give the best results. Comparing table 1 and 2, we see that random decompositions give about the same communication time whether the matrix to be decomposed is very structured as in A_i) or random as in A_{ii}). ORB, of course, does find the correlations in A_i) and give decompositions which have much lower communications in this case. The results in table 2 also allow us to improve the worst case estimates given in equations (7-I) and (8). We find that the use of the improving routing described in the appendix decreases (7-I) by a factor of approximately two and corresponding increases the speedup estimate in (8) by the same factor.

If we examine the "mistakes" made by ORB we see that certainly the subdomains are non-optimal as illustrated in figure 5(b). Further probably more important in examples studied here, ORB misaligns the bisected subgraphs and incorrectly assigns subdomains to processors. It is likely that one can improve the algorithms in section IV in this regard but in the following we speculate on a different approach.

We regard ORB as providing a coarse graining of the domain; in another language it divides the problem into *objects* which are to be manipulated separately. Considered as such, it is natural to use ORB to divide the graph (domain) into more *objects* than processors. A reasonable choice may be to divide the original domain into N_0 *objects* with perhaps:

$$N_0=(8-16)N \tag{32}$$

This easily achieved by using ORB as described in section IV but in a mode that corresponds to a hypercube of dimension 3 or 4 more than the original. Ref. [11] describes how these *objects* can be manipulated by a dynamic load balancer.

Now we fix the make up of the *objects* but allow their assignment to particular processors to change. This can nicely be obtained by using simulated annealing to improve the decomposition of the *objects*. Note that the original graph defined between elements defines a new graph among the *objects* into which the elements have been clustered. Thus the formalism of section III can be applied immediately. This approach has two significant advantages over simulated annealing or neural networks applied directly to elements.

(i) Simulated annealing and neural networks works speedily on small systems but they may be impractical on to apply it large $M \sim 10^6$ systems. It will definitely take too long a time to find a good decomposition for simulated annealing and neural nets may also be too time consuming. (26) implies a modest number ($\lesssim 10{,}000$ for the current practical limit $N \leq 1024$) of objects and a much faster annealing. The ORB step to reach (32) is deterministic and linear in system size and, for choices like $P_{cut}=16$, $P_{align}=0$, will be quite practical on very large systems.

(ii) Simulated annealing and neural networks applied to the basic elements can be trapped in local minima. It is not possible to use them to evolve from a *ring* or ORB generated decomposition as these represent local minima. To find a better global minimum it is necessary to "waste" the ORB step by melting the system (using large a temperature T) and re equilibrating it. What one really wants is in fact a coherent move of groups of elements which is proposed by the simulated annealing or neural network algorithms applied to *objects*.

The advantages that ORB has in forming *objects* or the initial decomposition do not extend to its use dynamically to redistribute objects of elements corresponding to a dynamically varying A. There are two useful comments here

(iii) Neural network or simulated annealing methods naturally run on the hypercube itself [9] and this is the only natural way to implement a dynamic load balancer. ORB has steps, like finding distances between points, which have concurrency but for which we have not developed a version that would run on the hypercube and so be efficiently useable by a load balancer.

(iv) Dynamic load balancing differs from the initial decomposition problem in that the matrix A is typically not dramatically different from its value when the current decomposition was performed. Thus, the current assignment of elements to processors should be reasonable. Under these circumstances the neural network method should take a time

$$T^{rebalance} \sim (\log_2 N)M \tag{33}$$

rather than

$$T^{initial\ balance} \sim (\log_2 N)M^{\alpha},\ \alpha = 1 + 1/d_s > 1 \tag{34}$$

The time (33) is now, unlike (34), competitive with ORB and as described in (iii) is speeded up by a factor N if implemented concurrently on the hypercube.

We intend to explore the hybrid ORB-simulated annealing/neural net approach in future research. We also hope to look at a wider variety of underlying matrices (graphs) A. Particle dynamics and neural simulation problems provide a rich source of such sparse matrices with an interconnect that has structure which is not trivially exploited by a simple domain decomposition. It would be interesting to find out if truly random matrices are ever important. It is interesting that ORB/simulated annealing/neural networks applied to any matrix A will in fact find any existing correlations! Maybe some problems that are currently considered random have hidden structure that can be uncovered by the techniques discussed in this paper.

Acknowledgements

I would like to thank Jon Flower for his help. Wojtek Furmanski developed the neural network approach reported here.

Appendix A: Routing on the Hypercube

Here we describe the algorithm used to route a given element x_i to its set of associated nodes $\{P_j\}$. These are nodes P_j such that A_{ji} is non zero. The algorithm used is as follows:

- Initialize list of visited nodes to P_i.
- Set route length to 0.
- ○ Loop over all visited nodes and all targeted nodes $\{P_j\}$ to be visited. Route x_i to target $P_{j_{min}}$ chosen so that total route length at this stage is minimized.
- If total route length $\geq N-1$, set route length = $N-1$ and end.
- If all target nodes visited, end.

Otherwise add $P_{j_{min}}$ and all intermediates nodes on route to $P_{j_{min}}$ to visited list and restart at ○.

This is a simple approximation to the solution of the traveling salesman on the hypercube. x_i is the salesman, $\{P_j\}$ are cities to be visited. It may be easy to improve the algorithm but it seemed to give reasonable results.

References

1. C^3P-385, "A Review of Automatic Load Balancing and Decomposition Methods for the Hypercube," G. C. Fox, November 1986
2. C^3P-255, "Concurrent Computation and the Theory of Complex Systems", G.C. Fox, S.W. Otto, March 3, 1986
3. C^3P-292, "A Preprocessor for Irregular Finite Element Problems", J.W. Flower, S.W. Otto, M.C. Salama, June 1986
4. C^3P-363, "Load Balancing by a Neural Network," G. C. Fox, W. Furmanski, September 1986
5. C^3P-314, "Optimal Communication Algorithms on the Hypercube", G.C. Fox, W. Furmanski, July 8, 1986
6. W. Furmanski, Private Communication, 1986
7. S. Kirkpatrick, C.D. Gelatt Jr., and M.P.Vecchi, Science 220, 671 (1983); S. Kirkpatrick, J. Stat Phys 34 , 975 (1984)
8. C^3P-214, "Monte Carlo Physics on a Concurrent Processor", G.C. Fox, S. W. Otto, E. A. Umland, November 6, 1985. Published in special issue of Journal of Statistical Physics, Vol. 43,1209, Plenum Press, 1986
9. J. J. Hopfield and D. W. Tank, "Computing With Neural Circuits: A Model," Science 233 , 625 (1986)
10. C^3P-371 "Optimization by a Computational Neural Net," R. D. Williams, October 10, 1986
11. C^3P-328, "The Implementation of a Dynamic Load Balancer", C. Fox, A. Kolawa, R. Williams, November 1986, submitted to 1986 Knoxville Hypercube Conference.

Table 1: Decompositions of Two Dimensional Meshes
Coefficients C in $T_{comm} = Ct_{comm}M$
(a) 16×16 Square Mesh

Decomposition N	Best	$P_{cut}=0$ $P_{align}=0$	$P_{cut}=1$ $P_{align}=0$	$P_{cut}=16$ $P_{align}=0$	$P_{cut}=16$ $P_{align}=1$	$P_{cut}=-1$ $P_{align}=1$	$P_{cut}=1$ $P_{align}=1$	Ring	Random
4	0.25	0.25	0.25	0.25	0.25	0.258	0.25	0.305	2.00
8	0.5	0.566	0.582	0.484	0.484	0.598	0.582	0.664	3.23
16	0.75	1.0	0.836	0.809	0.930	0.988	0.836	1.38	4.32
32	1.25	1.65	1.58	1.43	1.57	1.51	1.50	2.40	5.85
64	1.75	2.65	2.49	2.35	2.56	2.38	2.72	3.28	7.25

(b) 64×64 Square Mesh

Decomposition N	Best	$P_{cut}=0$ $P_{align}=0$	$P_{cut}=1$ $P_{align}=0$	$P_{cut}=16$ $P_{align}=0$	$P_{cut}=16$ $P_{align}=1$	$P_{cut}=-1$ $P_{align}=1$	$P_{cut}=1$ $P_{align}=1$	Ring	Random
4	0.0625	0.0625	0.065	0.0625	0.0625	0.0659	0.0625	0.076	2.09
8	0.125	0.131	0.142	0.140	0.140	0.156	0.142	0.162	3.33
16	0.188	0.216	0.230	0.225	0.225	0.229	0.229	0.333	4.66
32	0.375	0.385	0.477	0.446	0.434	0.426	0.449	0.671	6.03
64	0.438	0.640	0.671	0.619	0.649	0.615	0.673	1.35	7.45

(c) Irregular plate with 544 total mesh points studied in Reference 3

Decomposition N	Best	$P_{cut}=0$ $P_{align}=0$	$P_{cut}=1$ $P_{align}=0$	$P_{cut}=16$ $P_{align}=0$	$P_{cut}=16$ $P_{align}=1$	$P_{cut}=-1$ $P_{align}=1$	$P_{cut}=1$ $P_{align}=1$	Ring	Random
16	0.601	0.72	0.67	0.68	0.63	0.68	0.68	1.38	4.52

Table 2: Decompositions of Random Matrices with L=4
(number of nonzero columns per row)
Coefficients C in $T_{comm} = Ct_{comm}M$
(a) M=256 rows

Decom-position N	$P_{cut}=0$	$P_{cut}=1$ $P_{align}=0$	$P_{cut}=16$ $P_{align}=0$	$P_{cut}=16$ $P_{align}=1$	$P_{cut}=-1$ $P_{align}=1$	$P_{cut}=1$ $P_{align}=1$	Ring $P_{align}=1$	Random
4	1.77	1.74	1.77	1.73	1.65	1.74	1.90	2.12
8	2.87	2.77	2.72	2.74	2.62	2.77	3.05	3.31
16	4.05	3.87	3.76	3.77	3.75	3.97	4.29	4.68
32	5.28	5.10	5.00	4.95	4.89	5.08	5.60	6.02
64	6.58	6.34	6.46	6.28	6.07	6.46	6.95	7.53

(b) M=4096 rows

Decom-position N	$P_{cut}=0$	$P_{cut}=1$ $P_{align}=0$	$P_{cut}=16$ $P_{align}=0$	$P_{cut}=16$ $P_{align}=1$	$P_{cut}=-1$ $P_{align}=1$	$P_{cut}=1$ $P_{align}=1$	Ring	Random
4	1.80	1.78	1.75	1.74	1.67	1.77	1.92	2.10
8	2.87	2.82	2.74	2.75	2.63	2.81	3.06	3.34
16	4.02	3.92	3.83	3.81	3.69	3.90	4.32	4.70
32	5.21	5.06	4.97	4.99	4.78	5.06	5.64	6.12
64	6.45	6.27	6.17	6.19	5.94	6.28	7.02	7.60

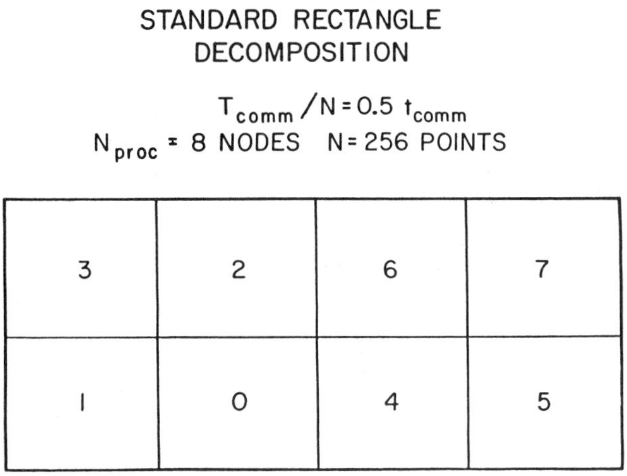

Figure 1

A Domain Decomposition for the case when A comes from the discretization of the two dimensional Laplacian. The cells assigned to a particular processor are labeled by its number.

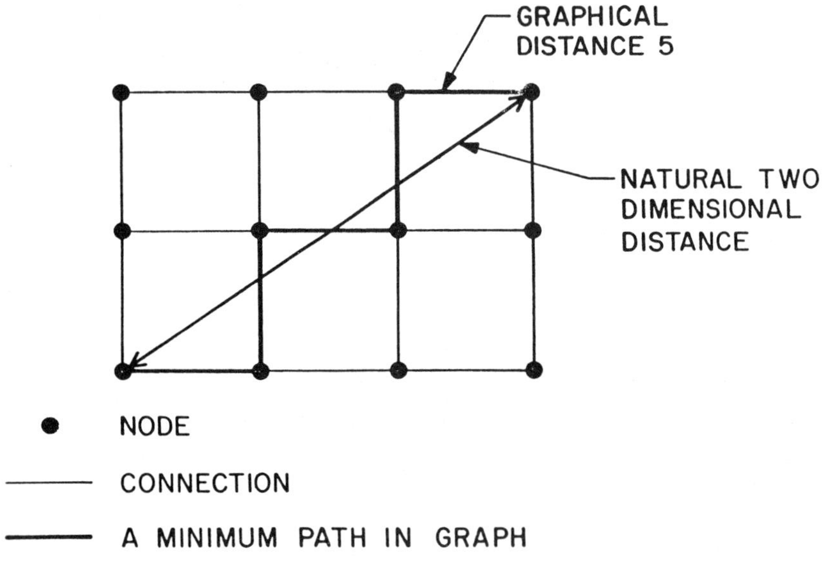

Figure 2

The distance d defined in equation (24) illustrated for the mesh problem.

RING DIVISION

$T_{comm}/N = 0.666\, t_{comm}$

N_{proc} = 8 NODES   N = 256 POINTS

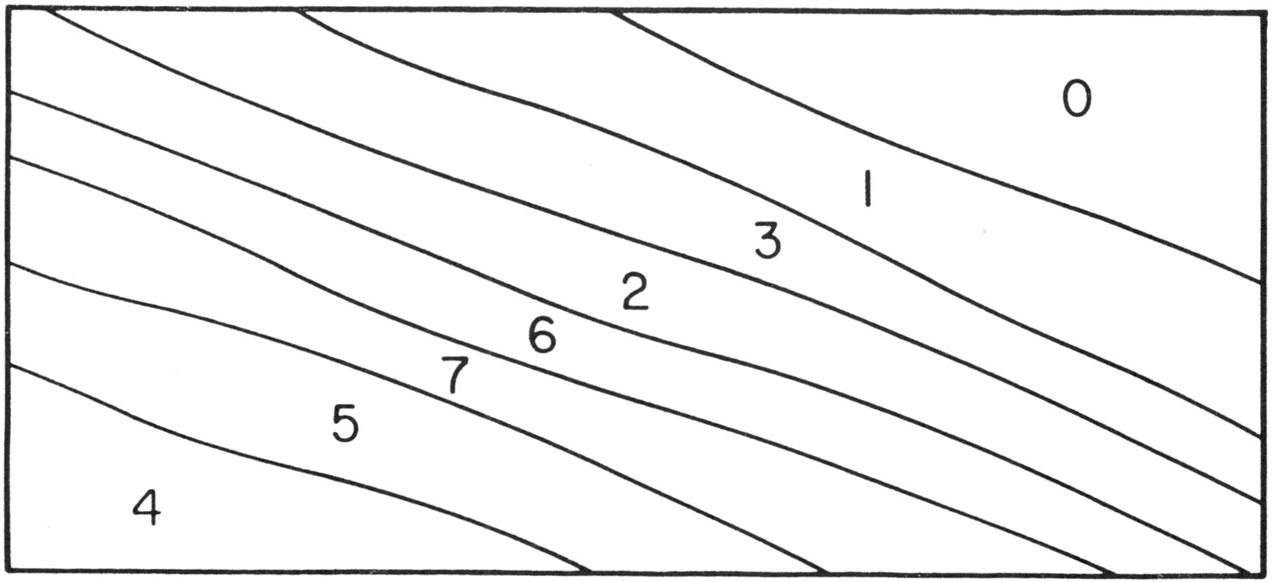

Figure 3

The *Ring* division algorithm described in section IV and illustrated for a 16x16 mesh decomposed over an 8 node hypercube.

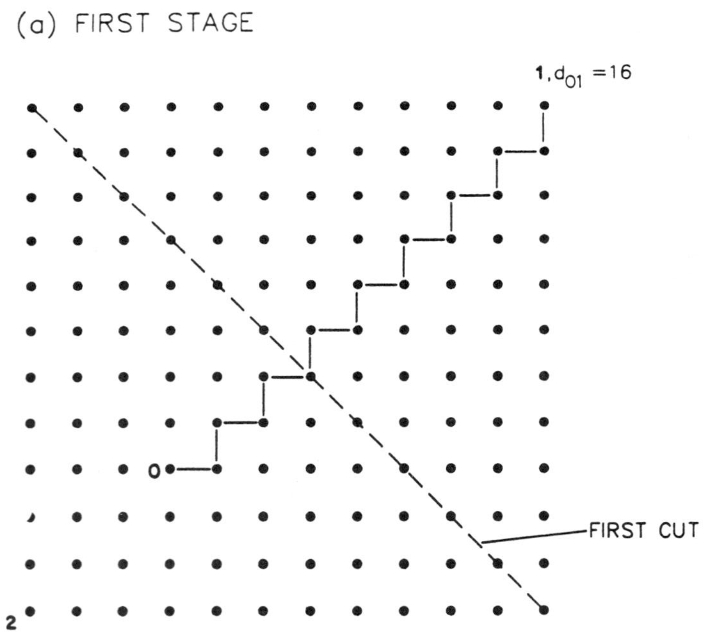

Figure 4a

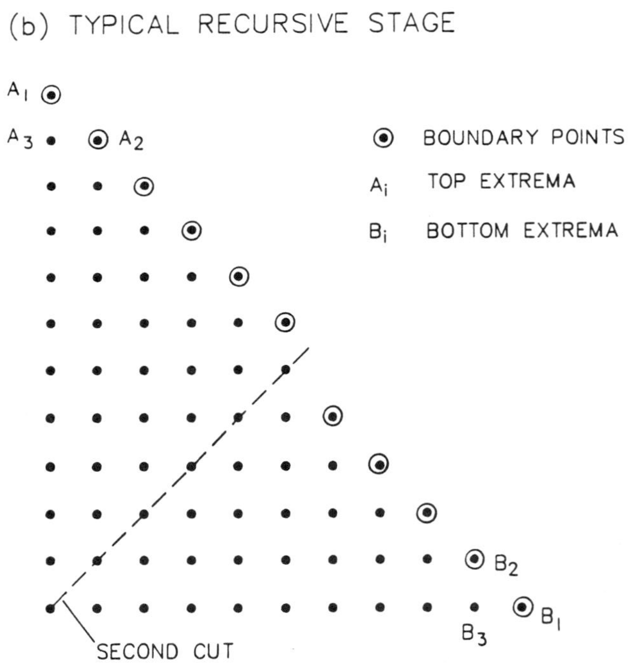

Figure 4b

Figure 4

Three stages of orthogonal recursive bisection illustrated for a simple two dimensional mesh. (Figures 4a, 4b, and 4c).

(c) TYPICAL ALIGNMENT STEP

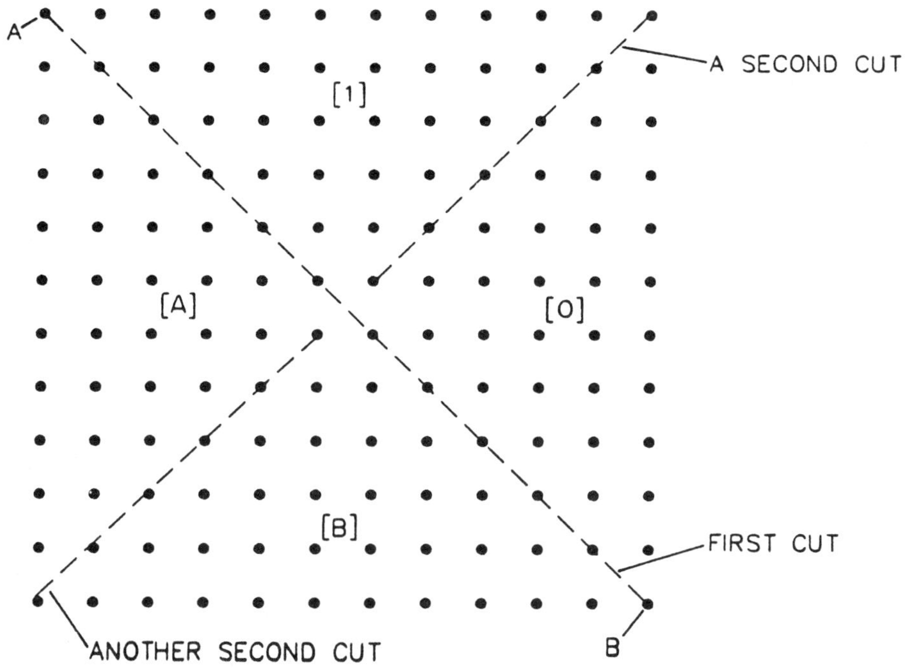

GRAPH HAS BEEN BISECTED INTO FOUR REGIONS
[0] [1] [A] [B]

Form $D_0 = d_R(B \text{ to } [0])$

$D_1 = d_R(B \text{ to } [1])$

$D_2 = d_R(A \text{ to } [0])$

$D_3 = d_R(A \text{ to } [1])$

If D_0 or D_3 smallest [B]=2, [A]=3

D_1 or D_2 smallest [B]=3, [A]=2

Figure 4c

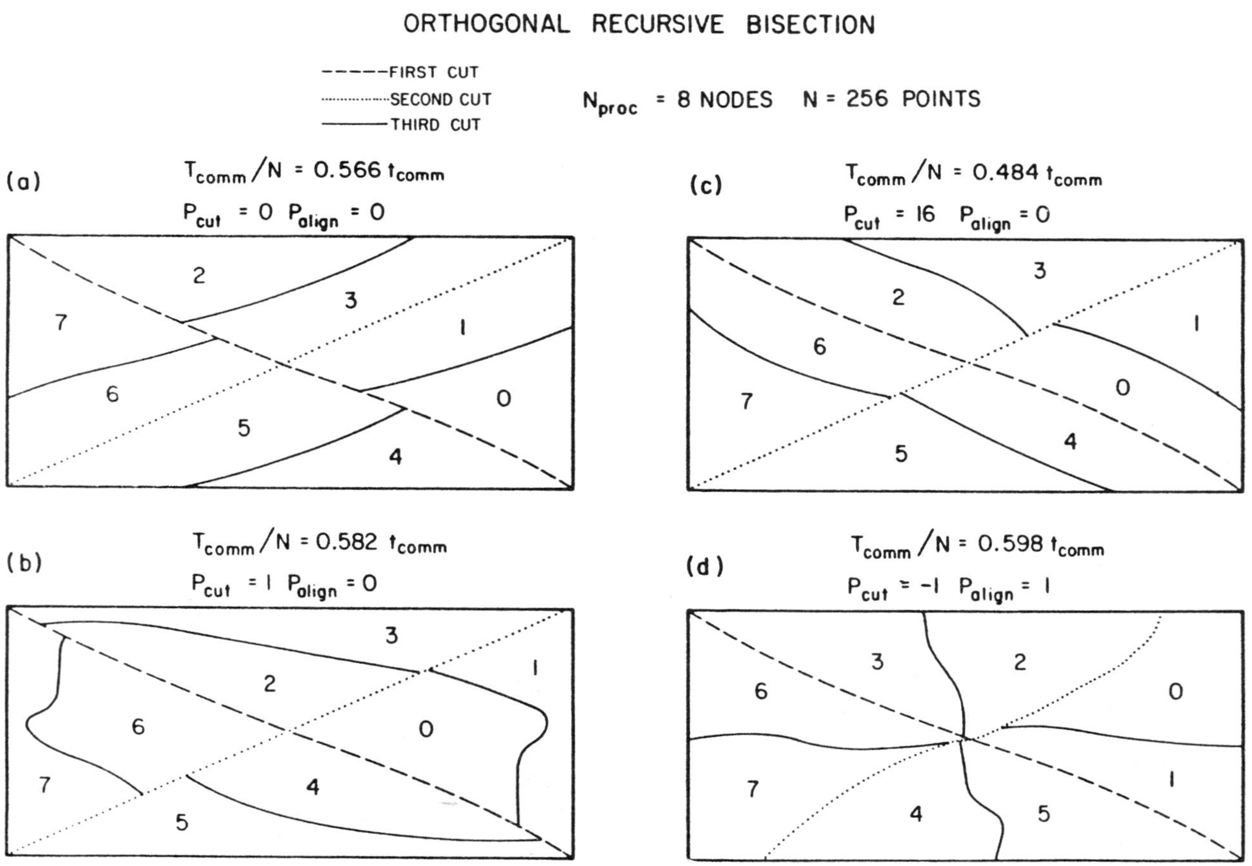

Figure 5

Some examples of decompositions generated by ORB for a 16x16 mesh decomposed onto 8 processors.

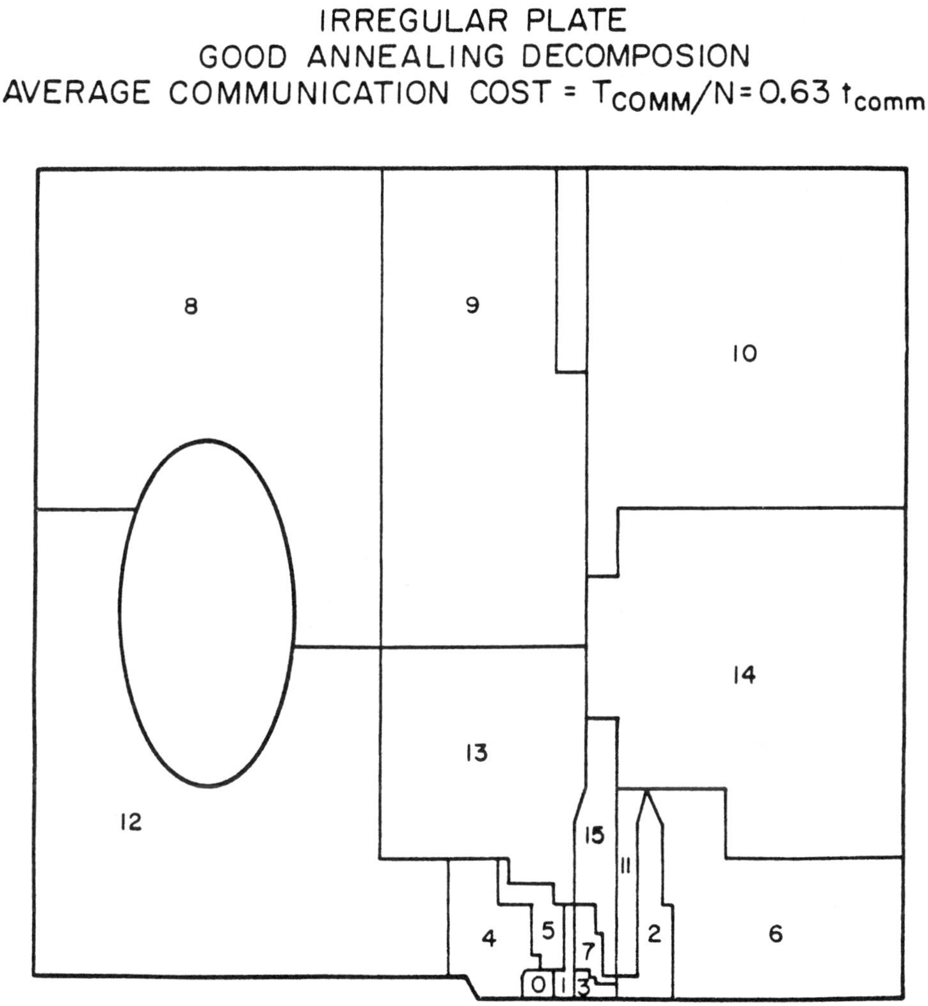

Figure 6

A good decomposition of the irregular plate problem studied in Ref. 3 and found by simulated annealing. This problem has 544 nodal points decomposed over 16 processors.

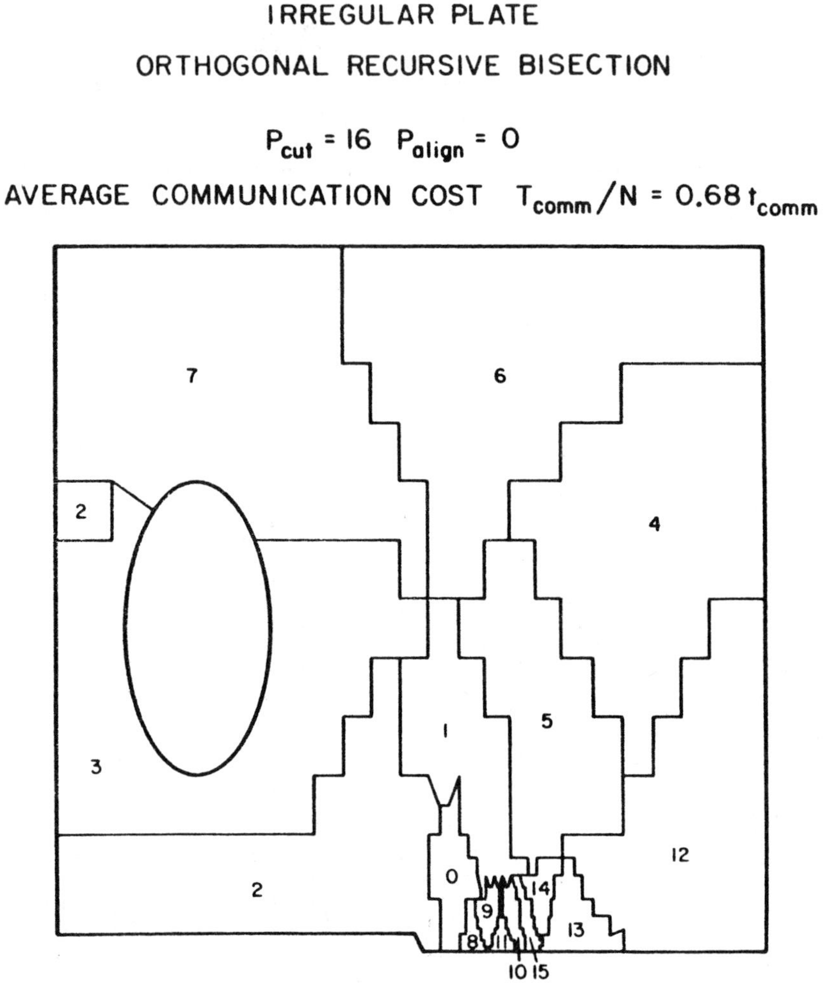

Figure 7

The same problem as shown in Figure 6 but now decomposed by the ORB technique with parameters $P_{cwt}=16$, $P_{align}=0$.

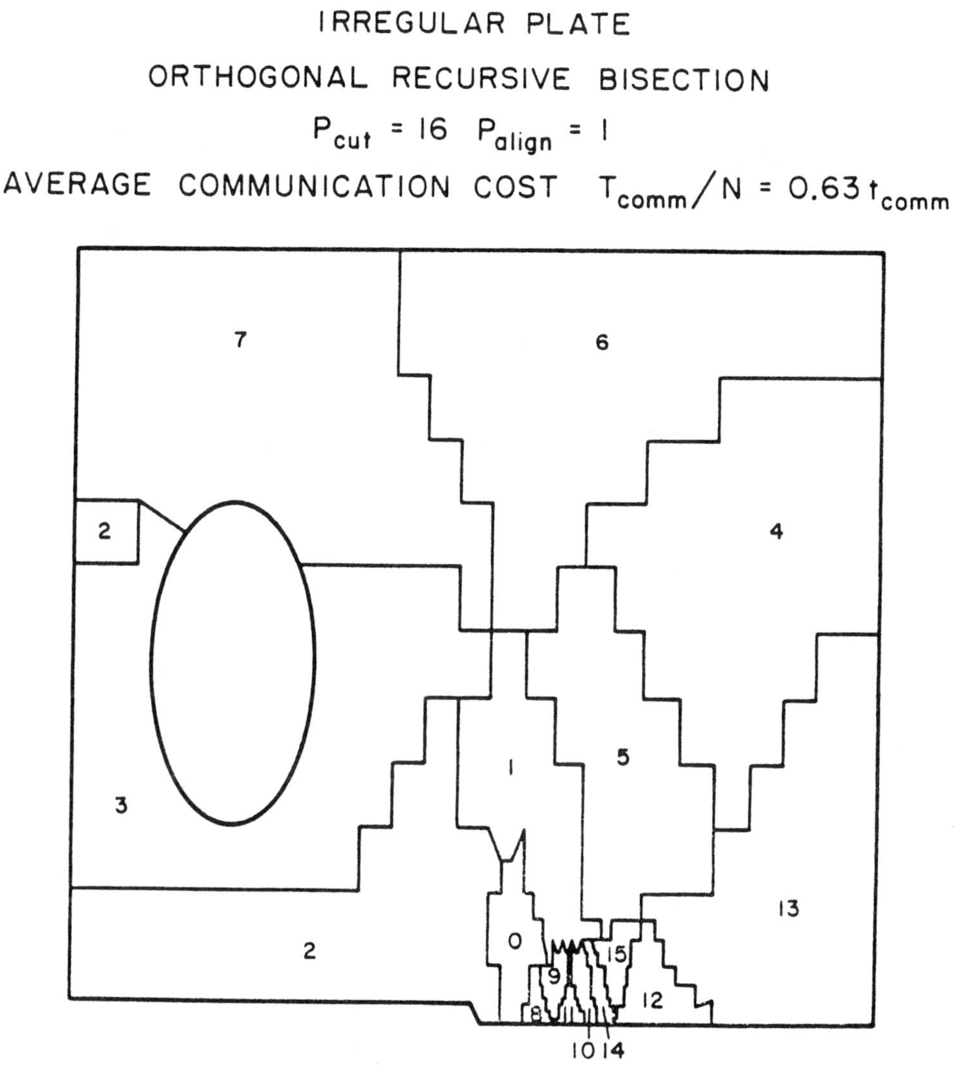

Figure 8

The same bisection as in Figure 7 but with the alignment parameter $P_{align} = 1$.

A Review of Automatic Load Balancing and Decomposition Methods for the Hypercube

Geoffrey C. Fox †
California Institute of Technology
Pasadena, CA 91125
December 5, 1986

Presented at Minnesota Institute for Mathematics and Its Applications Workshop November 6, 1986

Abstract

We give a brief review of the research at Caltech on methods to automatically decompose or load balance problems on the hypercube concurrent computer. We have several promising methods and are implementing a dynamic load balancer on the hypercube.

I: Introduction

At the Minnesota workshop, I gave a general review of the Caltech research on load balancing techniques for problems running on the hypercube. This report is a brief summary of this talk while as a companion contribution [1] to the proceedings, I consider one particular application, sparse matrix-vector multiplication, in detail. There, I show how to apply the simulated annealing and neural network methods and introduce and test a new graphical technique, orthogonal recursive bisection or ORB. Neither of these papers is complete but between them, they illustrate the Caltech research effort. This review does contain a complete set of references to our current reports.

In Section II, I present a general framework for the work formulated as an optimization problem in the context of a general theory of complex systems. In Section III, we review the various decomposition techniques while in IV, we discuss their use as a dynamic load balancer running side by side with the user code on the hypercube. Section V speculates on the future.

We will not distinguish the terms decomposition and load balancing here although the former tends to refer to static and the latter to dynamic implementations that change as the problem progresses. Most of the ideas discussed are equally applicable to either static or dynamic load balancing and, in general, we will discuss the latter as a more general scenario.

† Work supported in part by DOE grant DE-FG03-85ER25009, Parsons and System Development Foundations, and the office of the program manager of the Joint Tactical Fusion Office.

II: Complex Systems and Optimization

We have shown in Refs. 2 and 3 how the so called theory of complex systems is a useful framework for studying both load balancing and general properties of problems that determine their implementation on concurrent computers.

Crudely, we can consider a complex system as a collection of entities with some sort of static or dynamic connection between them. Some possibilities are illustrated in Figure 1. This shows that in many cases, one can characterize a complex system by a matrix A_{ij} where i,j label entities and A_{ij} is nonzero if and only if i and j are connected in the system. We will usually specialize to *spatially-connected* systems where the graph lies in a domain that can be decomposed over the hypercube. This should be contrasted with *causally-connected* systems such as the game tree found in computer chess [4]. In this case, the graph has directed links and in a hypercube implementation extends over time and not "space" - interpreted as the space formed by the nodes of the hypercube. We will usually restrict ourselves in this paper to spatially connected systems. These form a large class of important problems including circuit simulations, matrix algorithms, partial differential equations and particle dynamics. In these cases, either an iteration count or the simulation time naturally synchronizes the problem. The linkage between entities at a synchronization point defines the spatially connected graph. We will sometimes term this the *algorithmic graph* to distinguish it from the *dependency graph* defined by the linkage between the fundamental components in a (FORTRAN) program that implements this problem on a **sequential** computer.

Returning to Figure 1, we see that load balancing can be viewed as a graph partitioning task - we must divide the algorithmic graph up into N parts such that each part has the same size and the partitions cut the minimum number of links. Here we have $N = 2^{d_H}$ nodes in the hypercube and the two constraints correspond to equalizing the load on the nodes of the cube and to minimizing the communication between nodes. We can also view load balancing as minimizing the execution time of the problem on a concurrent machine. In either case, we see that decomposition or load balancing is "simply" an optimization problem and so can be tackled by any of your favorite optimization methods. As the optimization task is for all interesting problems, NP complete one does not try to find the optimal solution but rather one that is good enough. As we will see in the next section, our most powerful methods can be though of as mapping a complex system into a different system with the same graph. Then one uses the natural optimization method for this new system. This is summarized in Table 1 below.

Field	Entities	Optimization Method
Computer Science	Nodes of Abstract Graph	Heuristics Expert System
Physics	Particles	Simulated Annealing (Statistical Physics Monte Carlo)
Biology	Neurons	Neural Networks

Table 1: Three Optimization Methods

We can also draw interesting analogues with other resource allocation problems summarized in Table 2. Similar optimization methods to those described in Section III can be and in some cases have been applied to problems like those in Table 2.

Table 2: 10 Optimization Problems

1. Decomposing graphs (problems) onto hypercube
2. Coarse Graining: finding medium grain objects for object oriented approach
3. Fine grain dataflow assignment
4. Use of "caches" on shared memory concurrent machines and sequential computers
5. Optimizing compilers
6. Optimal adaptive meshes and determination of cells for "particle in the cell" problems.
7. Controlling Nation's Defense System
8. Designing Nation's Defense System
9. Controlling AT and T's phone network
10. Designing AT and T's phone network

III: Load Balancing Methods

Initially, we describe in IIIA and IIIB two heuristic methods which are easy to apply and work well in broad classes of problems. The final three subsection III C, D, E describe more powerful techniques that are generally applicable but require a more elaborate environment on the concurrent computer.

IIIA: The Scattered Decomposition

This is a simple method which sprinkles the nodes of the hypercube over the underlying data domain (or algorithmic graph). It is particularly useful in matrix problems which have no nearest neighbor structure and was introduced for this case by Fox [5] and its use is described in more detail by Refs 3, 6, and 7. It has also been shown valuable in irregular geometry finite element problems [8] and ray tracing algorithms [9].

In Ref [2], we show that the *scattered decomposition* is rigorously optimal for very rapidly varying or dynamic problems. We can use a general concept of *temperature* to quantify the concept of rapid variation by mapping a complex system into a gas of particles. The scattered decomposition corresponds to the high temperature limit.

IIIB: Self Scheduling

In our computer chess implementation on the hypercube , we treat the nodes of the concurrent computer as a set of "workers" to which processes are assigned as they become available. This is a conventional method of using a shared memory machine but can also be used in a distributed memory environment. In the case of chess, the "work" is a request for a node to evaluate a certain move. This request can recursively generate more requests as one progresses down the game tree. Our current implementation is rather rigid with a number of "master" nodes controlling servers. This leads to load imbalance (currently about a factor of 2) which is being improved by a more democratic method with "master" and "worker" functions being spread out over the nodes of the hypercube. We are developing a multi-tasking hypercube operating environment MOOOS that will make this easier to implement [10]. The chess strategy is the irregular version of our optimal algorithm to calculate several scalar products where the underlying vectors are decomposed over the hypercube [11]. Each scalar product becomes a tree and we spread the roots (the "masters") of these several trees evenly over the cube.

IIIC: Orthogonal Recursive Bisection

In the accompanying paper[1], we introduce a new graphical method for decomposing problems over the cube. This technique only uses the connection matrix A_{ij} discussed in Section II. We think of this as a graph with nodes labeled by i and links are formed between nodes i and j of the graph if and only if the corresponding matrix element A_{ij} is nonzero. In this picture, decomposition becomes a graph partitioning problem. We show in Ref. 1 how this can be done recursively, dividing the graph in two at each of $d_H = \log_2 N$ steps corresponding to the d_H dimensions of the hypercube. Here N is the total number of processors in the machine. This method only uses the natural distance $d(i,j)$ between any two points in the graph. We also introduce a concept of orthogonality for the divisions in successive dimensions.

This method, called ORB in Ref. [1], has the following important features:

- It is intrinsically approximate and has no natural way of improving the decomposition. However, as shown in Ref. [1], it gives good results for many problems.
- The decomposition time is given by

$$T^{ORB}_{decomp} \propto M \log_2 N \tag{1}$$

  where M is the total system size.
- It is not easy to implement a concurrent algorithm for ORB.

IIID: Simulated Annealing

The next two methods are analytic optimization methods which are based on minimizing an energy or objective function of the form[11,12,13]

$$E = E_1 + E_2 \tag{2}$$

Here E_1 is a measure of the load imbalance between the nodes and is typically taken as:

$$E_1 \propto \sum_{nodes} [\text{work per node}]^2 \tag{3}$$

while E_2 is a measure of the total communication. The exact form of E depends on details of hardware, e.g. are calculation and communication overlapped, which are unimportant here. The form (2) is certainly correct for the initial Caltech hypercube hardware and one can derive the correct relative normalization of E_1 and E_2 such that E is indeed the execution time on the concurrent machine [12, 20, 22]. In Equation (3), we have replaced the $\max_{nodes}$ (work per node) by the least squares sum. Near the maximum of E_1, this is unimportant in most problems. In fact, the form (2) has several advantages over the maximum of the work per node; in particular, (2) is a local function amenable to analytic optimization methods.

In Refs. [12] and [2], we show that the physics analogy mentioned earlier allows one to interpret E as the potential energy of a physical system whose particles are labeled by the index i. E_1 corresponds to a repulsive hard core and E_2 to an attractive long range potential. Minimizing E corresponds to finding the ground state of the system. The method of simulated annealing[15] just corresponds to the standard statistical physics approach to the determination of the ground state.

Figures 2 to 4 illustrate the application of this method to a simple finite element problem [2, 13]. In Figure 2, we show a finite element mesh generated by NASTRAN for the structural analysis of an irregular two dimensional plate. In Figure 3, we show that a simple equal area decomposition leads to severe load imbalance which is eliminated in Fig. 4 and corresponds to the application of the annealing algorithm to the initial imbalanced decomposition of Fig. 3.

We have shown in Ref. 2 that this physical analogy can lead to phase transitions corresponding to the onset of different optimal decompositions as a function of the system temperature mentioned earlier.

Simulated annealing has the following key features:

- The method is always approximate but with an appropriate annealing schedule can generate arbitrarily good solutions.
- It is very hard to estimate the decomposition time $T^{S.A.}_{decomp}$ which will depend drastically on the required goodness of the decomposition.
- The method can be efficiently implemented on a concurrent machine and will run with high efficiency on the hypercube. In Ref. [16], we show that it is not necessary to enforce the detailed balance constraint and this certainly makes the concurrent implementation much simpler.

IIIE: Neural Networks

Our most recent research has concentrated on the neural network approach to optimization introduced by Hopfield and Tank for the traveling salesman problem [17]. We show in Ref. 18, that neural nets are typically better than simulated annealing in many physics problems related to the study of spin glasses. The basic idea of neural networks is to find the minimum of the energy function E_1 defined in equation (2) by a deterministic set of equations. This is explained in detail in Ref. [19] and reviewed and compared with simulated annealing in Ref. 1. Although the biological analogy[17] suggests a general form of the neural net equations, there are many possible networks. In Ref. [19], we introduce the **bold** network and show it to be better than other algorithms for decomposing regular and irregular grid problems. We have since started to look at other topologies and have shown neural nets to be excellent for decomposing trees. In Figure 5, we show typical results corresponding to the decomposition of a mesh containing 1600 points. the histograms are averaged over 200 different initially random decompositions. The overhead is typically 50% greater than the optimal solution and this does not degrade significantly as one increases the size of the hypercube. One can show that the decomposition time,

$$T^{neural}_{decomp} \propto M^{1+1/d_s} . \log_2 N \tag{4}$$

where the system of size M has dimension d_s. The results in Fig. 5, correspond to $d_s = 2$ and a fixed number of $2M^{1/2}$ updates with each involving the update of the decomposition of all M points in the system. In Fig. 6, we compare neural nets with simulated annealing and show that, at least for the implementations of Refs. 19 and 13 respectively, the new method converges faster. The key features of the neural net method are:

- The method can be improved by increasing the number of iterations but one can get trapped in local minima.
- The method takes a "known" time given in (4) above.
- The evaluation of neural networks is similar to any circuit simulation and well suited to a straightforward concurrent implementation.

IV: A Dynamic Load Balancer

In Ref. 20, we review our current plans to implement a dynamic load balancer on the hypercube. The essential features of this are as follows:

- The load balancer and user program will co-exist on the hypercube running as separate sets of tasks on each node.
- Initially, we will only consider problems where the user (program) can estimate the energy function E - or equivalently the connection matrix A_{ij} and the calculation complexity C_i of each point.
- The balancer will not move individual entities (nodal points in Figures 2 to 4) but rather subdomains controlled by a complete process or task. These can be considered as *objects* [1, 20]. This will make the load balancing transparent to the user. Unfortunately, such a strategy is only possible on the hypercubes with memory management on each node, and this excludes many of our favorite machines.....
- The strategy for moving *objects* from processor to processor will be determined by a concurrent implementation of the simulated annealing or neural network methods.
- The initial decomposition into *objects* can often be performed by the user but the method of Section IIIC, orthogonal recursive bisection, is an attractive alternative [1].
- We have not yet thought through the issues connected with interplay of the load balancer and the user program i.e. what criteria should one use to invoke the balancer. We expect to explore some simple ideas with our initial implementations [20].

V: Conclusions

A year ago, we thought that automatic decomposition or load balancing was the key issue in determining the general applicability of hypercubes. To our surprise, we have found several methods that work well and the development of the algorithms in Section III is perhaps most limited by the lack of problems that are hard enough! We would like help in identifying the difficult problems and putting them in a form where we can study their decomposition. Complex event driven simulations and certain sparse matrix problems need study.

We believe that the dynamic load balancer reviewed in Section IV is technically hard to implement but that we understand the essential issues and the project described in Ref. 20, will be successful.

Even when the load balancer runs efficiently on the hypercube, the user must still write the code to control each *object*. We wonder if it may be possible to be more ambitions. Namely, one can formulate a more ambitious optimization than that studied in sections II and III. Can we take the dependency graph formed by the sequential code implementing a given problem and decompose this directly? This would both form objects, assign them to nodes of the cube **and** generate the necessary code to control them.

We believe that the neural network approach to optimization is very powerful and will be developed for many applications; especially those connected with intelligence - that existing now in man and that we wish to endow on future computers. Clearly, our rather modest application of neural networks to decomposition will evolve as the field does. In particular, we expect to improve the neural network decomposition time from $M^{1+1/d}$, to $M \log M$.

Acknowledgements

I would like to thank my many collaborators in this research - in particular D. Jefferson at UCLA and J. Flower, W. Furmanski, A. Kolawa, S. Otto, and R. Williams at Caltech. I also thank M. Schultz for inviting me to the interesting workshop.

References

1. "A Graphical Approach Load Balancing and Sparse Matrix Vector Multiplication on the Hypercube", Geoffrey C. Fox, Presented at Minnesota Institute for Mathematics and Its Applications Workshop, C3P-327B, November 6, 1986.

2. "Concurrent Computation and the Theory of Complex Systems", Geoffrey C. Fox, Steve W. Otto, March 3, 1986, submitted to proceedings of 1985 Hypercube Conference at Knoxville, C3P-255, August 1985.

3. "Solving Problems on Concurrent Processors", Geoffrey C. Fox, Mark A. Johnson, Gregory A. Lyzenga, Steve W. Otto, John K. Salmon, November 1, 1986, to be published by Prentice Hall.

4. "Chess on the Hypercube", E. Felten, R. Morison, S. Otto, K. Barish, R. Fatland, F. Ho, C3P-383, November 1986.

5. "Square Matrix Decomposition: Symmetric, Local, Scattered" G. Fox, C3P-97, August 13, 1984.

6. "A Banded Matrix LU Decomposition on the Hypercube", T. Aldcroft, A. Cisneros, G. Fox, W. Furmanski, D. Walker, C3P-348.

7. "Matrix", G. C. Fox and W. Furmanski, C3P-386.

8. "The Scattered Decomposition for Finite Elements", R. Morison, S. Otto, C3P-286, May 19, 1986.

9. "Graphics Ray Tracing and An Improved Hypercube Programming Environment", J. Salmon, C3P-345, July 31, 1986.

   "Static and Dynamic Database Distribution for Graphics Ray Tracing on the Hypercube", Proposal to JPL Directors Discretionary Fund, J. Goldsmith and J. Salmon, C3P-360.

10. "The MOOOS System Software", (Multitasking, object oriented), John Salmon, C3P-346, August 4, 1986.

11. "Optimal Communication Algorithms on the Hypercube", G. C. Fox and W. Furmanski, C3P-314, July 8, 1986.

    "Communication Algorithms for Regular Convolutions on the Hypercube", G. C. Fox, W. Furmanski, C3P-329, September 1, 1986, submitted to 1986 Knoxville Hypercube Conference.

12. "Monte Carlo Physics on a Concurrent Processor", G. C. Fox, S. W. Otto, E. A. Umland, C3P-214, November 6, 1985. Published in special issue of Journal of Statistical Physics, Vol. 43, 1209, Plenum Press, 1986, CALT-68-1315.

13. "A Preprocessor for Irregular Finite Element Problems", J. W. Flower, S. W. Otto, and M. C. Salama, C3P-292, June 1986.

14. G. Fox, D. Jefferson, S. Otto, Paper in Preparation.

15. S. Kirkpatrick, C. D. Gelatt Jr., and M. P. Vecchi, Science 220, 671 (1983); S. Kirkpatrick, J. Stat. Phys. 34, 975 (1984).

16. "Minimization by Simulated Annealing: Is Detailed Balance Necessary?" R. D. Williams, CALT-68-1407, C3P-354, September 3, 1986.

17. "Computing With Neural Circuits: A Model", J. J. Hopfield and D. W. Tank, Science 233, 625(1986).

18. "Optimization by a Computational Neural Net", R. D. Williams, C3P-371, October 10, 1986.

19. "Load Balancing by a Neural Network", G. C. Fox, and W. Furmanski, C3P-363, September 1986.

20. "The Implementation of a Dynamic Load Balancer", G. Fox, A. Kolawa, R. Williams, C3P-328, November 1986, submitted to 1986 Knoxville Hypercube Conference.

21. "Concurrent Processor Load Balancing as a Statistical Physics Problem", G. C. Fox and D. Jefferson, C3P-172, May 28, 1985.

22. G. Fox, unpublished note on load balancing, 1986.

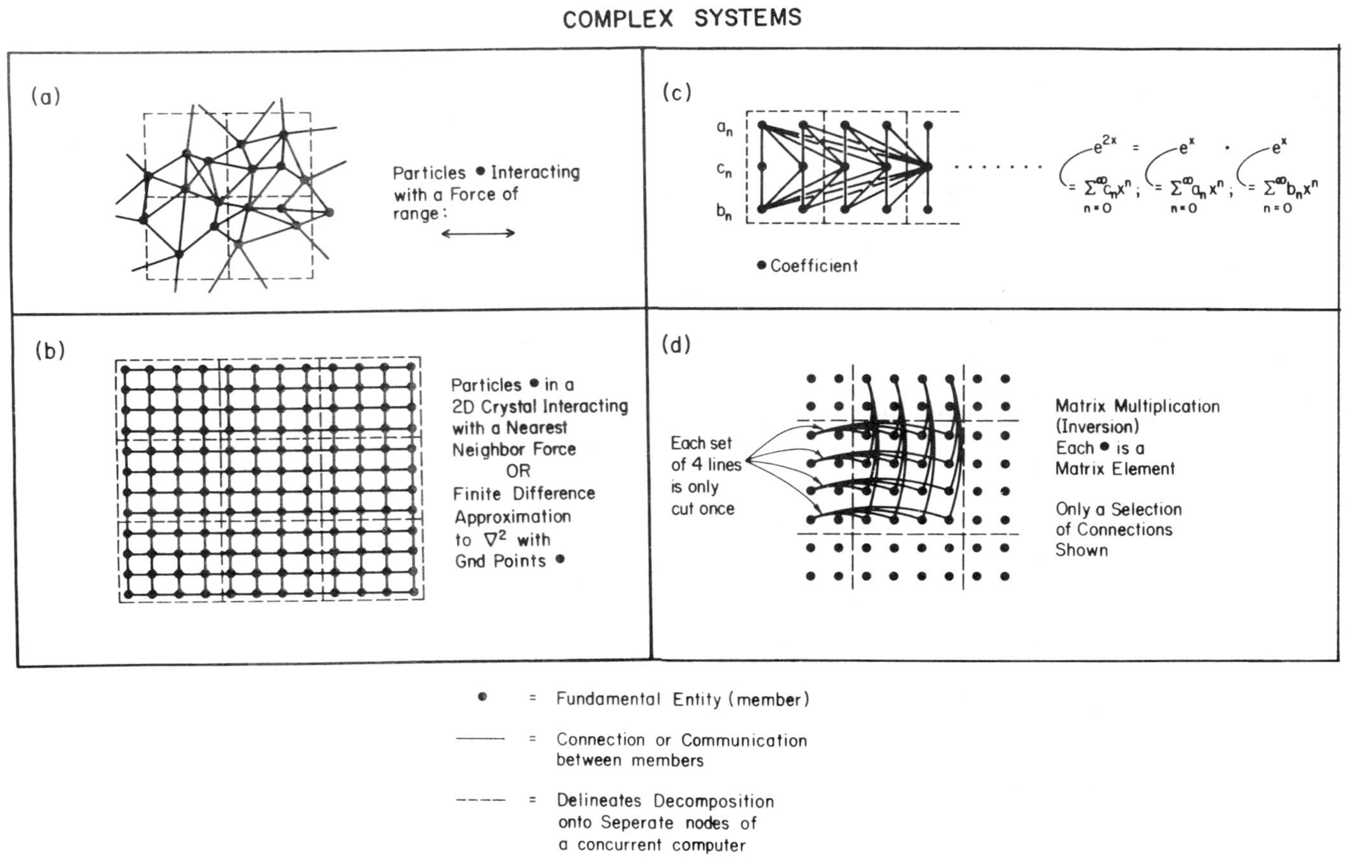

Figure 1

Four complex systems with their associated graphs with sample decompositions indicated by dashed lines.

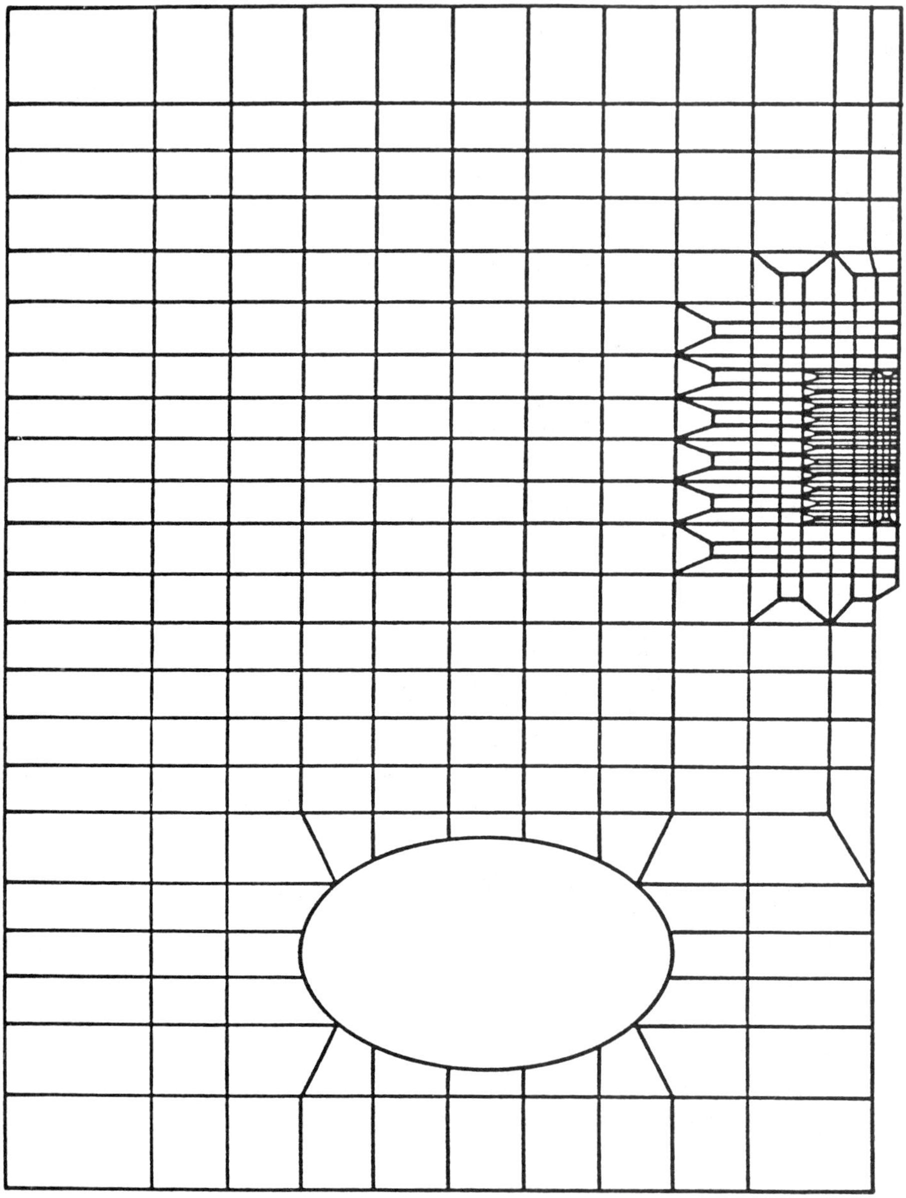

Figure 2

A 544 nodal point mesh generated by NASTRAN for a sample two dimensional plate finite element problem [13].

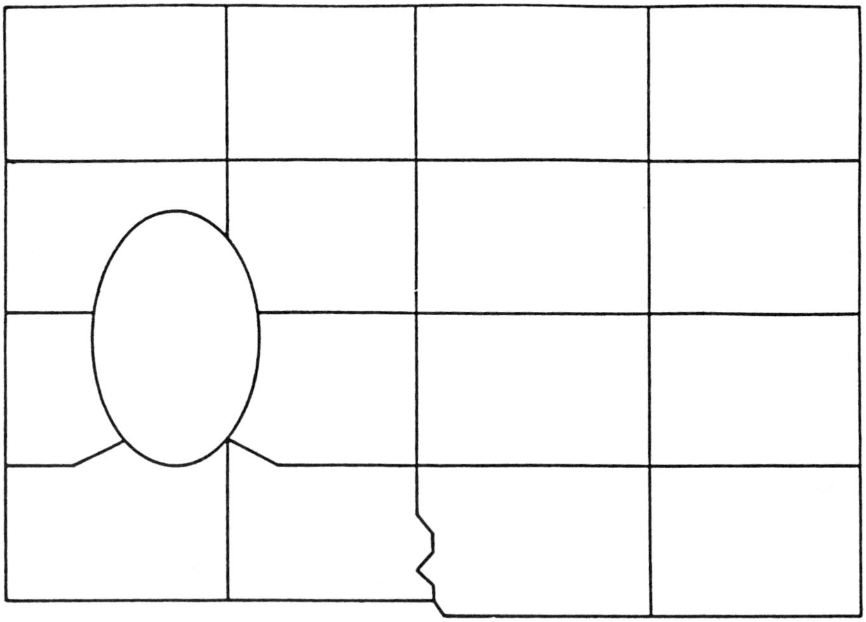

Figure 3a

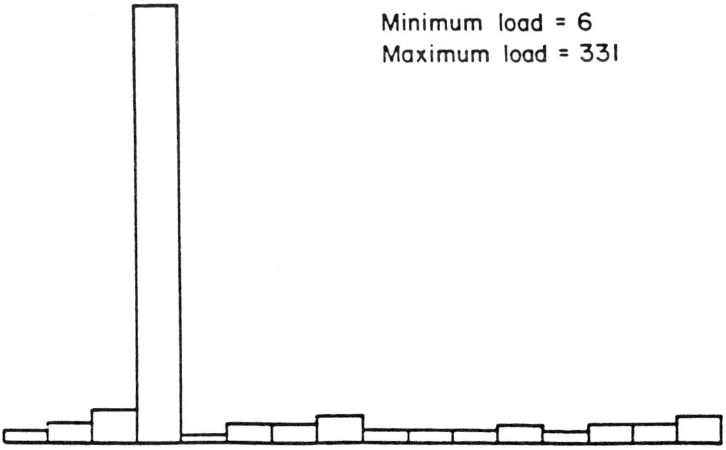

Figure 3b

Figure 3

A simple two dimensional equal area decomposition for the problem of Figure 2 showing severe load imbalance. This corresponds to a 16 node hypercube [13].

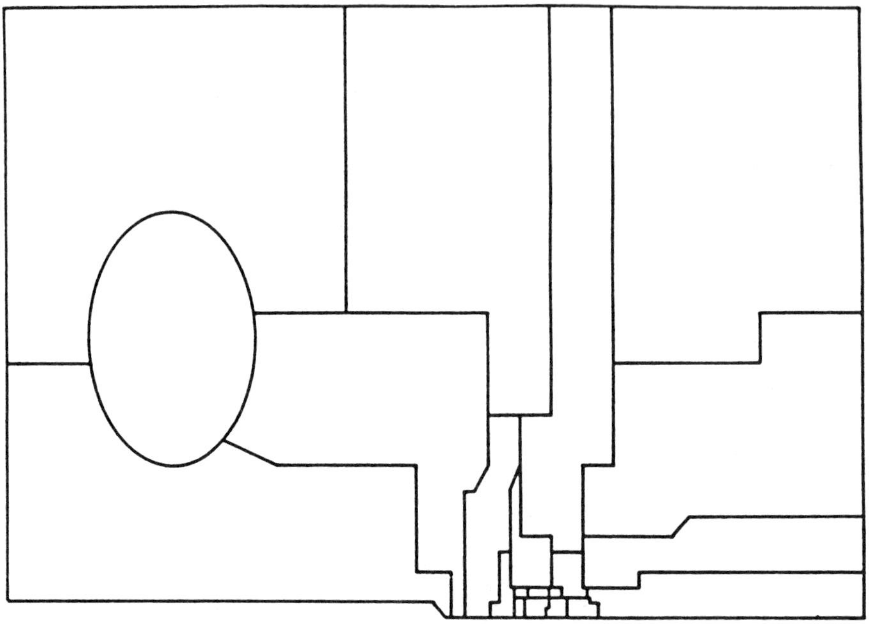

Figure 4a

Figure 4b

Figure 4

The result of simulating annealing applied to the problem of Figure 2 for decomposing onto a 16 node hypercube [13].

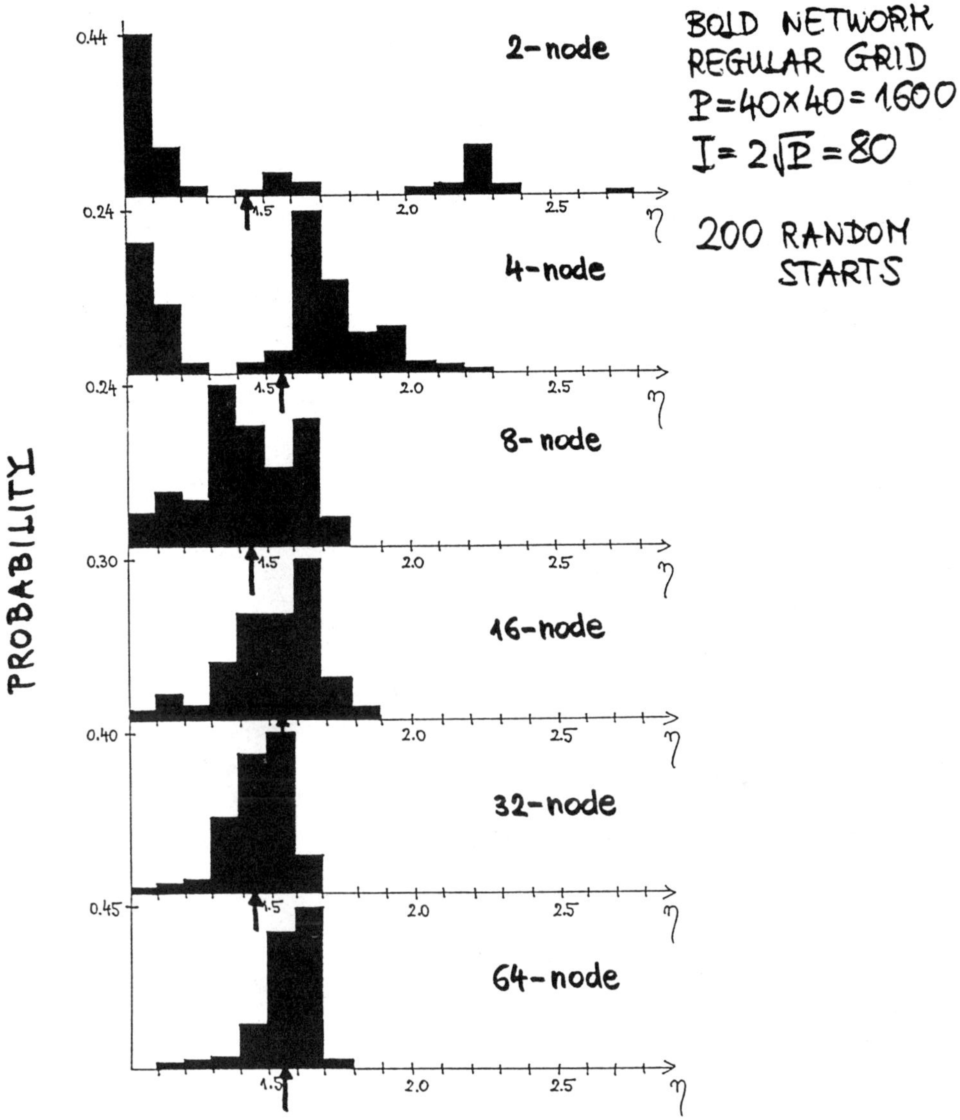

Figure 5

A plot of η - the ratio of actual communication cost to the optimal - for the *bold network* applied to a 40 x 40 grid averaged over 200 random initial conditions. The average values of $<\eta>$ are shown by arrows and results are given for 2, 4, 8, 16, 32, and 64 node hypercubes [19].

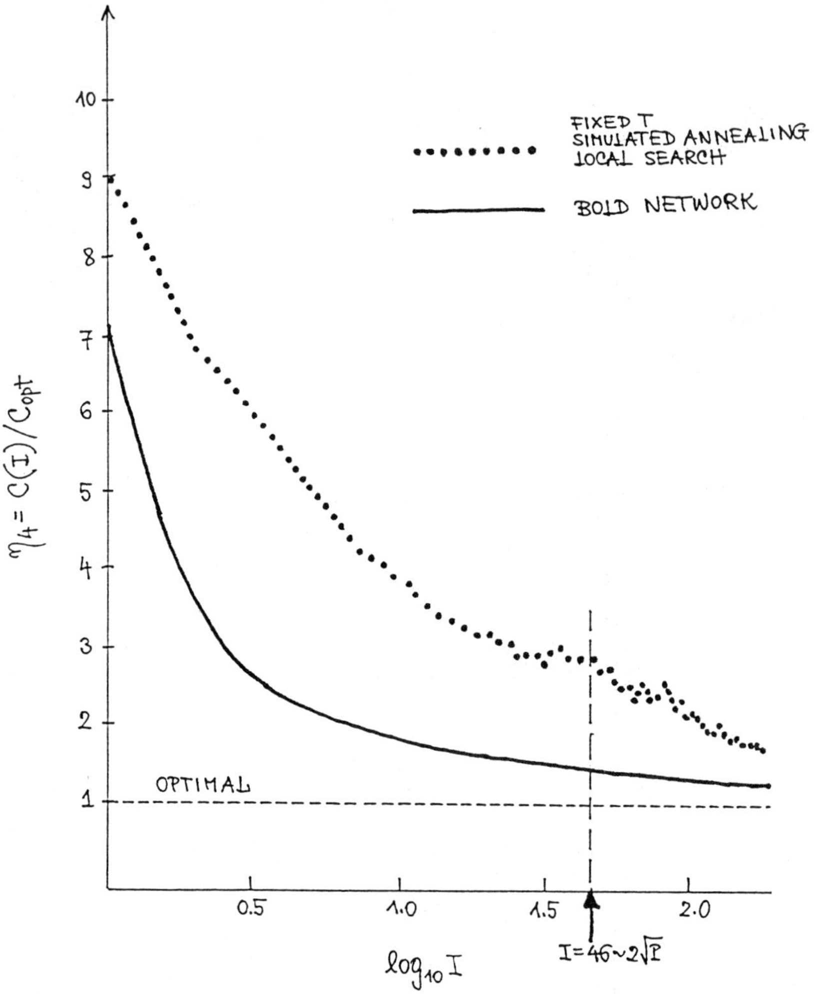

Figure 6

A plot of η - the ratio of actual communication cost to the optimal - for both simulated annealing [13] and the *bold neural network* [19] applied to the plate problem of Figure 2. η is plotted against an iteration count - each point is moved once in a single iteration.

ADAPTIVE METHODS FOR HYPERBOLIC PROBLEMS ON LOCAL MEMORY PARALLEL PROCESSORS

William Gropp
Yale University

1. Introduction

Parallel computers have opened up new directions for algorithm development. But the variety of parallel architectures and programming methods has lead to two apparently different models of computation. In this paper, I show that these models can be considered as special cases of memory access, similar to data cacheing. The numerical solution of hyperbolic partial differential equations is uses as an example, since simple algorithms for these problems exist and are representative of more realistic algorithms.

In any parallel algorithm, there are several considerations: the nature of the parallelism, the control of access to shared objects, and the programming tricks and optimizations. In many problems, the parallelism is expressed as a division of the problem by domains, either explicitly or through the division of a "do-loop" across multiple processors. Advantage is often taken of "locality of reference", where each processor works as much as possible with "local" data. This simplifies the control of access to shared data, a major problem in correctly programming parallel computers. Most algorithms (asynchronous ones excepted) must guarantee that data is ready before use. Finally, any number of tricks can be applied to speed the algorithm up. Some of these are general, such as noting that the algorithm itself may guarantee that data is ready before it is used; others are specific to a particular implementation or architecture.

Any parallel algorithm is concerned with the sharing of data in parallel. Only the parameters (costs) of doing so changes. We will discuss a unified model which suggests different techniques and rationals for various parallel architectures.

Adaptive methods for hyperbolic partial differential equations (PDEs) are a good model because there is a well defined locality of reference and obvious parallelism. Good solution techniques for, say, linear equations, aren't as obviously parallel (at least yet).

As is the case for vector supercomputers, a major constraint in any high performance parallel algorithm is "memory bandwidth". However, for a parallel computer, the "memory bandwidth" may contain message sending costs or critical section costs in addition to the usual memory hardware costs. The same techniques which have been applied to non-parallel, virtual memory computers may be appropriate for parallel computers.

1.1. Communication model

I will consider two types of parallel computer: message passing and shared memory with memory caches. Most parallel computers fit into one of these classes. In particular, most shared memory machines have caches, whether they are called that or not. Several shared memory machines have local memories (Cray 2, IBM RP3, BBN Butterfly), in each of these, access to the local memory is significantly faster than to the shared memory, making these local memories essentially caches.

In a message passing computer, information on remote data is transmitted by sending a message. This takes a time $s + rn$ for n words of data, where s is the startup time in seconds, and r is the transfer rate in seconds per word. Both s and r may depend on the relative position of the source and destination processors; if there isn't a direct connection between the two processors, a "multihop" message may be sent with increased cost

In a shared memory computer with a cache, there are several parameters. The most important for our purposes are the size of the cache C in words, the size of a cache page k, also in words, and the time to move a page into or out of the cache τ, in seconds per page.

An additional cost in shared memory computers is the time needed to establish a critical section or a barrier which insures that data is not used before it is available. This condition is automatically satisfied by message passing computers, since the message must be waited for before it can be received. I will ignore both of these costs in this paper. They are less important for the problems I'm considering here. However, for certain algorithms, such as Gaussian Elimination partitioned by rows on a ring, this idle time can be significant [1].

1.2. Overview of paper

This paper is divided into four parts. First, parallel programming techniques are discussed, showing that access to memory (or more abstractly, data objects) is one of the fundamental ideas. The relationship between shared memory and message passing is discussed.

Second, a discussion of simple, non-adaptive explicit methods is used to illustrate the similarities between message passing and shared memory, and how some of the same analysis can be used for each.

Next, fully adaptive algorithms are discussed, starting with their representation as a graph problem, and the parallel algorithm and programming issues that result. These issues include the granularity of a data item and load balancing costs.

I close with a discussion on how both message passing and shared memory can be considered as parallel programming analogues of different memory cacheing techniques, and what can be learned from those techniques.

2. Parallel programming concepts

Much discussion of parallel algorithms assumes either a shared memory or message passing model. For many algorithms, this is unnecessary; the concepts behind the algorithms are equally valid for both types of architectures. In this section I will discuss some of the considerations in creating a parallel algorithm and how they are independent of the parallel architecture.

In any creating a parallel algorithm, there are three basic considerations. First is to identify the parallelism. This may mean simply decomposing the problem domain of a conventional algorithm into p sections or it may mean an creating an entirely new algorithm. Other approaches used here include establishing a queue of tasks to be performed; each processor takes the next task from the queue.

The second step is to control the access to shared data items. This is essentially a *correctness* step; the algorithm must make sure that if a data item being computed by one processor is needed by another, it is not used before it is ready. It is here that the differences between the various hardware architectures is most apparent. In a shared memory computer, the most obvious way to share data is through shared memory, with control of access arrange either through the general mechanism of *critical sections*, or the simpler *barrier* (a barrier is simply a synchronization point which all processors must reach before any of them can continue). On a message passing computer, data is shared by explicitly passing it between processors, with access control handled implicitly by waiting for a message.

However, these are simply different ways of accomplishing the same goal. Which of these is more efficient, either in terms of programming or speed, depends on both the software and hardware tools available. Existing programming languages are not suited to the multiple threads of control present in a parallel algorithm, forcing the programmer to be very careful about the order of access of data. This suggests a programming method in which the access control is implied, such as message passing. However, of the existing hardware, the shared memory computers seem fastest, at least for moderate numbers of processors. Perhaps a middle ground, such as message passing software on shared memory hardware, is the best approach. Functional programming languages are taking an approach similar to this.

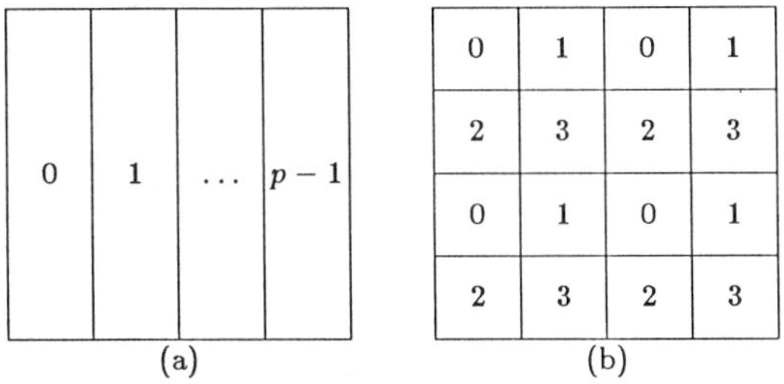

Figure 1: Two sample decompositions of a domain. (a) shows a 1-d division with one strip per processor. (b) shows a 2-d division with multiple pieces per processor. The labels are the processor numbers.

The third step is to optimize the algorithm for the parallel architecture. This might include taking advantage of shared memory, or changing algorithm parameters such as buffer sizes or order of operations to permit the maximum utilization of resources. Here, for example, access control steps in the algorithm may be identified as unnecessary or multiple copies of data created to reduce memory contention or message traffic. The point is that these operations are all *optimizations*, and are implementation details. As such, they should not be done prematurely nor without good reason. Programming parallel computers is difficult enough without the added complexity of unnecessary program optimizations.

3. Explicit non-adaptive algorithms

I start by considering the case of simple, explicit difference schemes for hyperbolic PDEs with no mesh adaptivity. I will show that there is a close relationship between messages and shared memory by considering them as different data cacheing schemes.

Explicit methods for hyperbolic equations are relatively simple. Essentially, the approximate solution at the next time step and at a point is formed by combining values of the solution in a small region about that point. Written more formally, for a hyperbolic partial differential equation in 2 space unknowns, and an approximation solution $A_{ij}(t_k)$, an explicit method can be written as

$$A_{ij}(t_{k+1}) = F(A_{ij}(t_k))$$

where the operator F is a combination of shift operators in space. As a very simple example, a (unstable) method might have

$$A_{ij}(t_{k+1}) = A_{ij}(t_k) + \frac{\Delta t}{2\Delta x}\left(A_{i+1,j}(t_k) - A_{i-1,j}(t_k) + A_{i,j+1}(t_k) - A_{i,j-1}(t_k)\right)$$

An important feature of such methods is the *locality of reference*: the new values of A depend only on nearby values of A at the previous time step. This is actually a reflection of a fundamental property of hyperbolic PDEs: causality.

3.1. Decomposition by domain

One of the most obvious approaches to dividing these problems among parallel processors is to divide the domain into regions, usually regular, then have each processor compute values for mesh points in its assigned region. Only along the boundaries of the domains do processors need information from other processors (I am assuming an explicit scheme with a simple stensil).

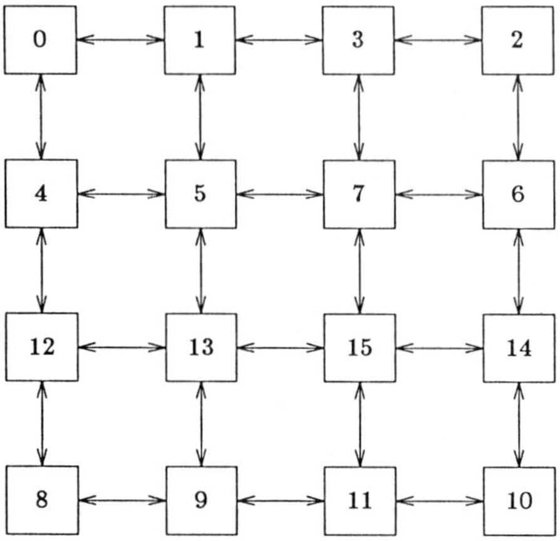

Figure 2: A 2-d mesh connected processor. Labels are the processors, arrows are the communication links. The labeling is a 2-d Gray code suitable for hypercubes.

A decomposition can be into 1, 2, or 3-dimensional slices, with one or more slices per processor. The choice of the best decomposition for a particular problem and computer depends on the topology of the computer (even if invisible to the programmer), the communication or memory access costs, the availability of resources such as memory locks, the need for load balancing, and the programming complexity.

Some sample decompositions are shown in Figure 1. The decomposition in Figure 1(a) is among the simplest, and is suitable for almost any parallel architecture. The decomposition in (b) is more suited for problems with an uneven work load over the domain, as each processor gets sections of the domain from many places.

In deciding which kind of decomposition to use, we must have a model of the costs of the various approaches. Since it turns out that, asymptotically in the size of the problem, the communication costs are negligible, we could choose any decomposition. But in reality, the constants involved in the communication terms can be large, and for realistic problems, the communication terms can be or comparable size to the "asymptotically dominant" terms.

3.2. Two examples

In this section I present two examples of estimates of communication costs; one for a mesh connected, message passing computer, and one for a completely connected, shared memory computer with caches.

3.2.1. Mesh connected message passing

I will consider a 2 dimensional problem, decomposed into either 1-d or 2-d slices, with one slice per processor. The parallel processor will have direct connections between the necessary neighbors. The division of processors with a 2-d decomposition of the domain and 16 processors, is shown in Figure 2.

Using the communication model from the introduction, it is easy to show that the cost for transfering data between neighboring processors is

$$\begin{aligned} C_{1d} &= 2(s + rn) \\ C_{2d} &= 4\left(s + \frac{rn}{\sqrt{p}}\right) \end{aligned}$$

where the domain is a square of side n, and there are p processors. From this, we can determine a number of facts. First, the 2-d decomposition, despite its better asymptotic behavior, is not always the best method. For example, if the ratio (for $\sqrt{p} \geq 3$)

$$\frac{s}{r} > n\left(1 - \frac{2}{\sqrt{p}}\right) > n,$$

then $C_{1d} < C_{2d}$. This in fact is the case for several commercial hypercubes and common values of n. However, the advantage of the 1-d decomposition is never more than a factor or two over the 2-d decomposition, as can be seen in the ratio

$$\begin{aligned} \frac{C_{1d}}{C_{2d}} &= \frac{1}{2}\frac{s + rn}{s + \frac{rn}{\sqrt{p}}} \\ &\geq \frac{1}{2}. \end{aligned}$$

Thus, everything else being equal, the best single choice of these for a robust program would be the 2-d decomposition.

3.2.2. Complete connection with cache

The parallel machine for this discussion is a shared memory computer with data caches on each processor. There is one cache per processor, and all memory references go through the cache.

In this case, the problem is with cache misses; if the cache never misses and the cache fill time is considered part of the memory cost, then there is no cost except for a barrier every time step to make sure one step is completed before the next begins. However, these are very big assumptions, and arranging for optimal cache performance is in many ways similar to optimizing message traffic.

First, consider a 1 dimensional decomposition and a 5 point stensil. If we are to avoid cache misses, then roughly 2 columns of data must be in the cache at any time. Specifically, if the explicit algorithm runs down columns of the domain, there must be at least

$$2\left(\frac{n}{k} + 1\right)$$

pages in the cache, with 2 pages read for every k points, assuming one word per mesh point, where each cache page holds k points. If the cache size C is smaller than this, then cache misses will occur, and performance will decline. Figure 3(a) shows diagrams this situation. Also notice that the Least Recently Used (LRU) algorithm will write out pages across rows instead of down columns, thus increasing the number of cache pages which need to be in the cache to prevent unnecessary cache misses. If the LRU algorithm is used to decide which cache pages to write out, then roughly $3n/k$ pages must fit into the cache.

If instead a 2 dimensional decomposition is used, then only

$$2\left(\frac{n}{k\sqrt{p}} + 1\right)$$

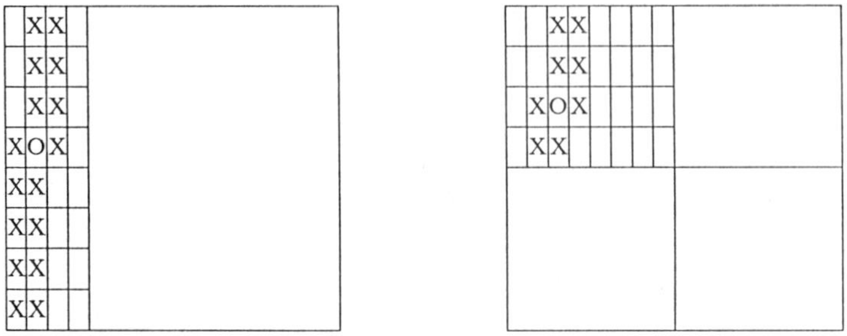

Figure 3: Cache pages in use of (a) 1-d slicing of the domain and (b) 2-d slicing. Only one processor is shown. An "O" marks the page at the center of the stensil, "X"s mark the pages needed for the current page or subsequent pages.

pages need to be in the cache at any one time. To compare these, note that if the cache size must be roughly $2n$ words in the 1-d decomposition and only $2n/\sqrt{p}$ in the 2-d decomposition.

In both cases, if the cache is too small, the number of cache pages that must be read roughly doubles. If the cache bandwidth is closely matched to the floating point speed, then this can effectively halve the speed of the program.

3.2.3. Comments

Despite the similarity between these two results, cache problems are not often considered. There are a number of possible reasons for this. One is that cache hardware has over the years been made very fast and efficient. The penalty for a cache miss in not necessarily large. Another reason is multiprogramming. With many users sharing the same processors, the cache is divided among them. Not only may the cost of a cache miss be shared among the users, but the available cache for each user may be only a fraction of available cache. Still, machines such as the Alliant FX-8, with their restricted cache bandwidth, can often reach their peak speed only with careful planning of the cache utilization (note that this machine has essentially a single cache shared among the processors, so the analysis presented here does not directly apply).

Beyond this, however, the considerations are quite similar. We can in fact look on message passing as predictive paging, made necessary by the expensive communications in the existing commercial machines. Further, the topology is important only as long as the communication time is long; this is true for both message passing and shared memory. As message passing matures, perhaps the (hardware) penalties in its use will be reduced. The various CalTech hypercubes [3] are examples of low cost message passing parallel computers. As a final comment, in a shared memory environment with caches, cache coherence is a potential problem; for example, Li [2] encountered some of these limits in his experiments with a shared memory system built on an Apollo ring.

4. Adaptive algorithms

Adaptive algorithms, at the programming level, are best viewed as graph management problems. Each node of the graph represents some work to be done, and each edge represents dependencies between the nodes. Dividing such a program up among multiple processors amounts to mapping the nodes of the graph to the processors.

This mapping raises a number of questions. The obvious ones are how to map the nodes to preserve data locality. That is, in a message passing computer, how to reduce the number of messages sent and the distance they must travel, and in a shared memory computer, how to reduce the shared memory traffic by keeping data in local memory or caches.

A less obvious question is what is the best decomposition of the problem into a graph in the first place. In particular, a large graph, where each node contains a single unknown or mesh point, may involve a great deal of "pointer chasing" and remote or shared memory references. If each node of the graph contains more work, then less time will be spent finding work to do. However, there is a penalty: as the nodes get larger it becomes harder to keep every processor busy. Further, in making the nodes larger, we may actually introduce more work, taking advantage of the lower overhead per mesh point. As the nodes get larger, these penalties make the larger nodes less and less attractive.

We can quantify some of these considerations with a simple example. Consider as a domain a grid of $n \times n$ mesh points, divided into squares of side b. Each of these squares is a node in our graph, which for the moment will be the trivial representation of a uniform mesh. Each of these nodes contains a $b \times b$ square submesh. By hypothesis, only the boundaries of these nodes must be transmitted to the neighboring processors. If we are using a message passing computer, the cost per processor is

$$C = \frac{4n^2}{pb^2}(s + rb),$$

since there must be n^2/p mesh points per processor and hence n^2/pb^2 blocks per processor.

We would like to pick b to minimize this cost. However, it is easy to see that

$$\frac{dC}{db} = -\frac{4n^2}{pb^2}\left(\frac{2s}{b} + r\right),$$

which is non-positive. Thus the cost decreases as b increases, with the minimum at the end point of the valid range of b, that is, at 1 block per processor ($b = n/\sqrt{p}$).

One contending effect is *load balancing.* If we now take this mesh, and if for some reason a roughly square area of size $R^2 = fn^2$, $0 < f < 1$ has most of the load (due to mesh refinement, non-linear solution iterations, etc.), then we want all of the processors to share in this load. If we use the decomposition in Figure 1(b), then we must have the blocks small enough to give each processor a block in this region of high load. This gives us the condition

$$\frac{R^2}{pb^2} \geq 1.$$

For example, if we expect roughly 10% of the region to generate most of the load, then the optimal b is

$$b \approx \frac{1}{3}\frac{n}{\sqrt{p}}.$$

What does all this suggest for adaptive algorithms? First, there are a number of contending effects. Large blocks minimize the data sharing overhead, whether it be message or cache traffic or accessing a critical section. Small blocks minimize the work load imbalance, modulo the load expended doing the load balancing. Perhaps the best way to deal with this is to use block algorithms, where node in the graph is a block of mesh points.

Second, each processor may have *read* access to the same data. On a local memory computer, several processors may have read-only copies of the same data. For example, in a sparse matrix computation, several processors may have copies of the same information on the lists in the sparse matrix representation. The additional overhead in checking to see if the data is local may be far smaller than always fetching it from global memory or some other processor's cache or local memory.

As in cacheing, the minimum amount of information which should be sent every time a message is sent should be proportional to the ratio of startup time to transfer rate. In particular, it takes only twice as long to send s/r words as it takes to send 0 words. Thus, in a message passing computer with high message startup time, only long messages should be sent. In many cases, by sending both the requested data and "nearby" data, the number of messages can be reduced. This is the principle on which conventional single processor cacheing systems depend; parallel algorithms and machines can take advantage of it as well.

We can also look at current message passing techniques as a form of predictive paging, which in practice has rarely been worth the additional effort it entails. For parallel algorithms, the delivery of data predictively should be considered an optimization, albeit a very valuable and often necessary one.

The point of these comments is to point out that message passing and shared memory are not that different, either in format or potential performance. The big difference is who is responsible for managing the access to remote objects, and how is it done.

References

[1] D. J. Baxter and W. D. Gropp, *Some Pipelined algorithms for the solution of dense linear systems on hypercubes,* 1987. in preparation.

[2] Kai Li, *Shared Virtual Memory on Loosely Coupled Multiprocessors,* Technical Report YALE/DCS/RR-492, Yale University, Department of Computer Science, September 1986.

[3] C. L. Seitz, *The Cosmic Cube,* Communications of the ACM, 28/1 (1985), pp. 22–33.

SYSTOLIC ALGORITHMS FOR THE PARALLEL SOLUTION OF DENSE SYMMETRIC POSITIVE-DEFINITE TOEPLITZ SYSTEMS

ILSE C.F. IPSEN

Department of Computer Science

Yale University

P.O. Box 2158, Yale Station

New Haven, CT 06520, USA

Abstract. The most popular method for the solution of linear systems of equations with Toeplitz coefficient matrix on a single processor is Levinson's algorithm, whose intermediate vectors form the Cholesky factor of the inverse of the Toeplitz matrix. However, Levinson's method is not amenable to efficient parallel implementation. In contrast, use of the Schur algorithm, whose intermediate vectors form the Cholesky factor of the Toeplitz matrix proper, makes it possible to perform the entire solution procedure on one processor array in time linear in the order of the matrix.

By means of the Levinson recursions we will show that all three phases of the Toeplitz system solution process: factorisation, forward elimination and backsubstitution, can be based on Schur recursions. This increased exploitation of the Toeplitz structure then leads to more efficient parallel implementations on systolic arrays.

Introduction

The aim of this paper is to discuss parallel methods for the solution of linear systems of equations

$$T_n x = f$$

whose coefficient matrices T_n are dense symmetric positive-definite Toeplitz matrices.

A symmetric Toeplitz matrix $T_n = (t_{kl})_{0 \le k,l \le n}$ of order $n+1$ is a matrix whose elements are constant along each diagonal, $t_{kl} = t_{|k-l|}$. The solution of a general, $n \times n$ system of equations by a direct method requires $O(n^3)$ operations. Since a $n \times n$ Toeplitz matrix is characterised by $O(n)$ rather than $O(n^2)$ parameters efficient algorithms for the solution of Toeplitz systems exhibit an operation count that is considerably smaller: the classical Levinson and Schur algorithms require $O(n^2)$ operations [1, 14, 15, 18] while the doubling algorithms require $O(n \log^2 n)$, cf. the early references [2, 9]. A thorough treatment of Toeplitz matrices is given in [10, 11], and a brief summary can be found in [17]. Numerical aspects of algorithms for Toeplitz matrices are reviewed in [5].

Development of parallel implementations for the solution of dense Toeplitz systems was motivated by the need to execute certain signal processing tasks in real-time. The preferred architectures are ***systolic arrays***, special-purpose devices built with Very Large Scale Integrated (VLSI) circuit technology [13]. Systolic arrays are homogeneous networks of tightly coupled, highly synchronised, simple processors that essentially operate in SIMD (Single Instruction Multiple Data stream) mode. Due to the repetitiveness of the computations and the regularity of the data dependencies systolic implementations can be described by means of linear transformations: the processor in which a quantity $r_{i,j}$ is computed as well as the time of its computation is expressed as a linear function in the indices i and j [7, 8]. To keep the approach simple and intuitive, implementation details will be omitted in this paper, they can be found in [7, 8].

Three classes of systolic arrays will be presented whose efficiency improves with increased exploitation of the Toeplitz structure in various phases of the solution process.

The classical method of choice for solving a $n \times n$ symmetric positive-definite Toeplitz system on a single processor is the Levinson algorithm [14]. The intermediate vectors generated by the Levinson algorithm form the Cholesky factor of the inverse of the Toeplitz matrix. Due to a sequence of n inner products, however, the lower bound of the parallel solution time on n processors is $O(n \log n)$.

The second classical method is the Schur algorithm [1, 15, 18]; its intermediate vectors form the Cholesky factor of the Toeplitz matrix. Although its operation count is fifty percent higher than that of Levinson's method, it is more amenable to parallel implementation: an array of $O(n)$ processors can determine the Cholesky factor of an order-n Toeplitz matrix in time $O(n)$ [3, 4, 6, 7, 8, 12, 16].

The solution to the Toeplitz system can be found by performing forward elimination with the transpose of the Cholesky factor and subsequent backsubstitution involving the Cholesky factor.

This necessitates the additional use of arrays for triangular system solution and intermediate storage of order $O(n^2)$ for the Cholesky factor during forward elimination [12, 16].

Instead of performing the usual forward elimination, recursions similar to the 'Schur recursions' in the factorisation may can be employed to modify the right-hand side vector, thus making it possible to employ the same type of array for factorisation and forward elimination. Intermediate storage of the Cholesky factor till the start of backsubstitution may be avoided by re-generating it on the fly [3, 4].

A final improvement in efficiency is achieved by also performing backsubstitution by Schur recursions. In this case it is possible to perform the whole solution process on one n-processor array in time $O(n)$ [7, 8].

Since it appears impossible to conceive systolic implementations of doubling algorihms, it can be concluded that the most efficient method hitherto to solve Toeplitz systems on systolic arrays is one that makes maximum use of Schur recursions.

Notation

A symmetric Toeplitz matrix of order $n+1$ will be denoted by T_n, where

$$T_n = \begin{pmatrix} t_0 & t_1 & \dots & \dots & t_n \\ t_1 & t_0 & \ddots & & \vdots \\ \vdots & \ddots & \ddots & \ddots & \vdots \\ \vdots & & \ddots & t_0 & t_1 \\ t_n & \dots & \dots & t_1 & t_0 \end{pmatrix},$$

and the sequence $t_i \dots t_k$ of Toeplitz matrix elements for $i \leq k$ will be denoted by $t_{i:k}$. Frequent use will be made of the fact that the first and last columns of T_n have the respective representations $t_{0:n}$ and $Jt_{0:n}$ where $J = (e_n \ \dots \ e_1 \ e_0)$ represents the 'exchange' matrix with ones on the antidiagonal, and e_i is a $(n+1) \times 1$ vector with a one in position i and zeros everywhere else, $0 \leq i \leq n$. At last, 0_k stands for the $k \times 1$ vector consisting of k zero elements; when $k = 0$ it is the empty vector.

A symmetric Toeplitz matrix T_n is centro-symmetric, that is

$$T_n = JT_nJ.$$

Direct methods with operation count $O(n^2)$ or less for the solution of dense Toeplitz systems of order n exploit this property and the resulting recursive nesting of Toeplitz matrices:

$$\left(\begin{array}{ccc|c} & & & t_n \\ & T_{n-1} & & \vdots \\ & & & t_1 \\ \hline t_n & \dots & t_1 & t_0 \end{array}\right) = T_n = JT_nJ = \left(\begin{array}{c|ccc} t_0 & t_1 & \dots & t_n \\ \hline t_1 & & & \\ \vdots & & T_{n-1} & \\ t_n & & & \end{array}\right).$$

The system $T_nx = f$ can be solved by recursively solving a system involving T_{n-1}.

The Levinson Algorithm

Levinson's algorithm computes the Cholesky factorisation of the inverse of a $n \times n$ Toeplitz matrices with $O(n^2)$ operations; the following derivation is partly based on the one given by Trench in [19]. Suppose the Cholesky factor L_n of $T_n^{-1} = L_n^T D_n^{-1} L_n$ is known, where L_n is unit lower triangular and D_n is diagonal (as T_n is symmetric positive-definite the diagonal elements of D_n are strictly positive). The Toeplitz matrix T_{n+1} of next higher order can be partitioned into a 2×2 block matrix as

$$T_{n+1} = \begin{pmatrix} T_n & t \\ t^T & t_0 \end{pmatrix}, \qquad t = Jt_{1:n+1} = (\, t_{n+1} \quad \dots \quad t_1 \,).$$

Solving

$$T_{n+1}^{-1} T_{n+1} = \begin{pmatrix} I_{n+1} & 0 \\ 0 & 1 \end{pmatrix},$$

where I_{n+1} is the identity matrix of order $n+1$, results in a 2×2 block partitioning for T_{n+1}^{-1}

$$T_{n+1}^{-1} = \begin{pmatrix} T_n^{-1} + T_n^{-1} t t^T T_n^{-1}/d & -T_n^{-1} t/d \\ -t^T T_n^{-1}/d & 1/d \end{pmatrix}.$$

Setting $\psi = -T_n^{-1} t$ yields

$$\begin{aligned} T_{n+1}^{-1} &= \begin{pmatrix} T_n^{-1} + \psi\psi^T/d & \psi/d \\ \psi^T/d & 1/d \end{pmatrix} = \begin{pmatrix} I_n & \psi \\ & 1 \end{pmatrix} \begin{pmatrix} T_n^{-1} & \\ & d^{-1} \end{pmatrix} \begin{pmatrix} I_n & \\ \psi^T & 1 \end{pmatrix} \\ &= \begin{pmatrix} L_n^T & \psi \\ & 1 \end{pmatrix} \begin{pmatrix} D_n^{-1} & \\ & d^{-1} \end{pmatrix} \begin{pmatrix} L_n & \\ \psi^T & 1 \end{pmatrix} = L_{n+1}^T D_{n+1}^{-1} L_{n+1}. \end{aligned} \tag{1}$$

Thus, $(\,\psi^T \quad 1\,)$ is the trailing row of the Cholesky factor L_{n+1} of T_{n+1}^{-1}. It remains to show that ψ and d can be computed in $O(n)$ steps.

To this end, suppose that the Cholesky factorisation of T_n^{-1} is already known:

$$T_n^{-1} = L_n^T D_n^{-1} L_n = \begin{pmatrix} T_{n-1}^{-1} + \psi_n \psi_n^T / d_n & \psi_n / d_n \\ \psi_n^T / d_n & 1/d_n \end{pmatrix}, \qquad D_n = \begin{pmatrix} d_0 & & \\ & \ddots & \\ & & d_n \end{pmatrix},$$

where $d_0 = t_0$ and ψ_n is a $n \times 1$ vector. The symmetry-centro symmetry of Toeplitz matrices implies for the inverse of the next higher-order matrix T_{n+1} that $T_{n+1}^{-1} = J T_{n+1}^{-1} J$, or in block form

$$\begin{pmatrix} T_n^{-1} + \psi_{n+1}\psi_{n+1}^t / d_{n+1} & \psi_{n+1}/d_{n+1} \\ \psi_{n+1}^T / d_{n+1} & 1/d_{n+1} \end{pmatrix} = \begin{pmatrix} 1/d_{n+1} & \psi_{n+1}^T J / d_{n+1} \\ J\psi_{n+1}/d_{n+1} & T_n^{-1} + J\psi_{n+1}\psi_{n+1}^t J / d_{n+1} \end{pmatrix}. \tag{2}$$

This gives for the trailing row $\psi_{n+1} = -T_n^{-1} J t_{1:n+1}$ of L_{n+1} the block form

$$\begin{pmatrix} -1/d_n & -\psi_n^T J / d_n \\ -J\psi_n / d_n & -T_{n-1}^{-1} - J\psi_n\psi_n^T J / d_n \end{pmatrix} \begin{pmatrix} t_{n+1} \\ J t_{1:n} \end{pmatrix} = \begin{pmatrix} -(t_{n+1} + \psi_n^T t_{1:n})/d_n \\ -T_{n-1}^{-1} J t_{1:n} - J\psi_n (t_{n+1} + \psi_n^T t_{1:n})/d_n \end{pmatrix}.$$

Denoting the term in brackets by

$$\rho_{n+1} = -(t_{n+1} + \psi_n^T t_{1:n})/d_n = -\begin{pmatrix} \psi_n^T & 1 \end{pmatrix} t_{1:n+1}/d_n, \qquad \rho_1 = -t_1/t_0, \tag{3}$$

and observing that $\psi_n = -T_{n-1}^{-1} J t_{1:n}$ gives

$$\psi_{n+1} = \begin{pmatrix} \rho_{n+1} \\ \psi_n + \rho_{n+1} J \psi_n \end{pmatrix}.$$

Consequently, with ψ_0 the empty vector, the trailing row of L_{n+1} can be obtained from the trailing row of L_n via

$$\begin{pmatrix} \psi_{n+1} \\ 1 \end{pmatrix} = \begin{pmatrix} 0 \\ \psi_n \\ 1 \end{pmatrix} + \rho_{n+1} \begin{pmatrix} 1 \\ J\psi_n \\ 0 \end{pmatrix}. \tag{4}$$

Remembering that d_n^{-1} is the bottom right element of T_n^{-1} and ρ_{n+1} is the leading element of ψ_{n+1} one gets with (2) for the bottom right element of T_{n+1}^{-1}

$$d_{n+1}^{-1} = d_n^{-1} + \rho_{n+1}^2 d_{n+1}^{-1} \quad \text{or} \quad d_{n+1} = d_n(1 - \rho_{n+1}^2), \qquad d_0 = t_0. \tag{5}$$

Note that the original paper by Levinson [14] does not contain the simple recursive computation of d_{n+1} from d_n and ρ_{n+1}.

The Levinson Algorithm

The Levinson algorithm computes the lower triangular Cholesky factor L_n of T_n^{-1} with kth row given by $(\psi_{k,0} \; \ldots \; \psi_{k,k-1} \; 1 \; 0_{n-k}^T)$.

$$d_0 = t_0$$

$$1 \le k \le n, \quad \rho_k = -(t_k + \sum_{j=1}^{k-1} \psi_{k-1,j-1} t_j)/d_{k-1}, \quad d_k = d_{k-1}(1-\rho_k^2), \quad \psi_{k,0} = \rho_k$$

$$(\psi_{k,0} \; \ldots \; \psi_{k,k-1}) = (1 \quad \rho_k) \begin{pmatrix} \psi_{k-1,0} & \ldots & \psi_{k-1,k-2} \\ \psi_{k-1,k-2} & \ldots & \psi_{k-1,0} \end{pmatrix}$$

The reason why Levinson's algorithm has little potential for parallelisation is that the vector ψ_n enters into the computation of ψ_{n+1} from *both* ends: in a linear combination of ψ_n and $J\psi_n$. Even if one were to maintain two separate copies, ψ_n and $J\psi_n$, the weight ρ_{n+1} in this combination would still depend on the *entire* vector ψ_n. Thus, it is not possible to pipeline successive recursions, and the lower bound on the time of a $n \times n$ parallel Cholesky factorisation of is $O(n \log n)$.

The Schur Algorithm

Since the inner-product (3) in the formation of ρ_{n+1} is the culprit for the poor parallel performance of Levinson's algorithm one could try to reformulate the algorithm so as to obviate the need for an explicit inner-product computation. This is accomplished by observing that (1) implies $d_n = (\psi_n^T \quad 1) J t_{0:n}$, and substituting this into (3) leads to

$$\rho_{n+1} = -\frac{(\psi_n^T \quad 1)\, t_{1:n+1}}{(\psi_n^T \quad 1)\, J t_{0:n}}.$$

Thus, the coefficient ρ_{n+1} is the ratio of two quantities that are obtained by multiplying $(\psi_n^T \quad 1)$ and its reverse by a column of the Toeplitz matrix. This is the basis for the so-called Schur algorithm [1, 15, 18], it avoids the inner-product by recursively 'updating' matrix-vector products involving the Toeplitz matrix, so that ρ_{n+1} can be formed as the ratio of two vector elements.

Unlike the Levinson algorithm which determines the Cholesky factor of the inverse, the Schur algorithm determines the Cholesky decomposition of the matrix proper. If the result of Levinson's method is $T_n^{-1} = L_n^T D_n^{-1} L_n$, where L_n is lower triangular, then the uniqueness of the Cholesky decomposition implies that $L_n T_n = D_n L_n^{-T}$ must be a scaled version of the upper triangular Cholesky factor U_n of T_n: $T_n = U_n^T D_n U_n$. Therefore it follows that $(\,(\,\psi_k^T \quad 1\,) \quad 0_{n-k}^T\,)\,T_n$ is the kth row of the scaled Cholesky factor $D_n U_n$ of T_n. Let 0_k^T represent a row vector of k zeros, then the kth row of $D_n U_n$ has the form

$$(\,(\,\psi_k^T \quad 1\,) \quad 0_{n-k}^T\,)\,T_n = (\,(\,\psi_k^T \quad 1\,) \quad 0_{n-k-1}^T \quad 0\,)\begin{pmatrix} T_k & A_k & * \\ A_k^T & T_{n-k-2} & * \\ * & * & t_0 \end{pmatrix}$$

$$= (\,(\,\psi_k^T \quad 1\,)\,T_k \quad (\,\psi_k^T \quad 1\,)\,A_k \quad *\,) = (\,0_k^T \quad d_k \quad (\,\psi_k^T \quad 1\,)\,A_k \quad *\,)$$

$$= (\,0_k^T \quad d_k \quad \nu_{k,0} \quad \ldots \quad \nu_{k,n-k-1}\,)\,, \tag{6}$$

where $*$ denotes unimportant terms and from (1)

$$(\,\psi_k^T \quad 1\,)\,T_k = (\,0_k^T \quad d_k\,)\,. \tag{7}$$

If it is possible to compute the non-zero elements $(\,d_k \quad \nu_{k,0} \quad \ldots \quad \nu_{k,n-k-1}\,)$ of the kth row of $D_n U_n$ with $O(n-k)$ operations then the Cholesky factorisation $T_n = U_n^T D_n U_n$ can be computed with $O(n^2)$ operations. It is now shown how to compute the vector of $\nu_{k,j}$ as a linear combination of two vectors by making use of the Levinson recursion as follows:

$$(\,(\,\psi_{k+1}^T \quad 1\,) \quad 0_{n-k-1}^T\,)\,T_n = (\,0 \quad (\,\psi_k^T \quad 1\,) \quad 0_{n-k-1}^T\,)\,T_n + \rho_{k+1}\,(\,(\,\psi_k^T \quad 1\,)\,J \quad 0 \quad 0_{n-k-1}^T\,)\,T_n.$$

With (3), (7) and (6) the first summand evaluates to

$$(\,0 \quad (\,\psi_k^T \quad 1\,) \quad 0_{n-k-1}^T\,)\begin{pmatrix} t_0 & t_{1:k+1}^T & * \\ t_{1:k+1} & T_k & A_k \\ * & A_k^T & T_{n-k-2} \end{pmatrix}$$

$$= (\,(\,\psi_k^T \quad 1\,)\,t_{1:k+1} \quad (\,\psi_k^T \quad 1\,)\,T_k \quad (\,\psi_k^T \quad 1\,)\,A_k\,) = (\,-\rho_{k+1} d_k \quad 0_k^T \quad d_k \quad (\,\psi_k^T \quad 1\,)\,A_k\,)$$

$$= (\,-\rho_{k+1} d_k \quad 0_k^T \quad d_k \quad \nu_{k,0} \quad \ldots \quad \nu_{k,n-k-2}\,)\,.$$

The second summand amounts to

$$\begin{pmatrix}(\psi_k^T & 1)J & 0 & 0_{n-k-1}^T\end{pmatrix}\begin{pmatrix} T_k & Jt_{1:k+1} & B_k \\ t_{1:k+1}^T J & t_0 & * \\ B_k^T & * & T_{n-k-2}\end{pmatrix}$$

$$= \begin{pmatrix}(\psi_k^T & 1)JT_k & (\psi_k^T & 1)t_{1:k+1} & (\psi_k^T & 1)JB_k\end{pmatrix} = \begin{pmatrix}d_k & 0_k^T & -\rho_{k+1}d_k & (\psi_k^T & 1)JB_k\end{pmatrix}$$

$$= \begin{pmatrix}d_k & 0_k^T & \mu_{k,0} & \ldots & \mu_{k,n-k-1}\end{pmatrix}.$$

where the leading element of the non-zero part is $\mu_{k,0} = -\rho_{k+1}d_k$.

Forming the linear combination of the two summands yields

$$\begin{aligned}&\begin{pmatrix}(\psi_{k+1}^T & 1) & 0_{n-k-1}^T\end{pmatrix}T_n \\ &\quad - \begin{pmatrix}-\rho_{k+1}d_k & 0_{n-k}^T & d_k & \nu_{k,0} & \ldots & \nu_{k,n-k-2}\end{pmatrix} + \rho_{k+1}\begin{pmatrix}d_k & 0_{n-k}^T & \mu_{k,0} & \ldots & \mu_{k,n-k-1}\end{pmatrix} \\ &\quad = \begin{pmatrix}0_{k+1}^T & d_{k+1} & \nu_{k+1,0} & \ldots & \nu_{k+1,n-(k+1)-1}\end{pmatrix}\end{aligned}$$

where due to (5) $d_{k+1} = d_k(1-\rho_{k+1}^2)$. Similarly, determination of the new vector elements $\mu_{k+1,i}$ is accomplished using the row-reversed version of (4)

$$\begin{aligned}&\begin{pmatrix}(\psi_{k+1}^T & 1)J & 0_{n-k-1}^T\end{pmatrix}T_n \\ &\quad = \begin{pmatrix}(\psi_k^T & 1)J & 0 & 0_{n-k-1}^T\end{pmatrix}T_n + \rho_{k+1}\begin{pmatrix}0 & (\psi_k^T & 1) & 0_{n-k-1}^T\end{pmatrix}T_n \\ &\quad = \begin{pmatrix}d_k & 0_k^T & \mu_{k,0} & \ldots & \mu_{k,n-k-1}\end{pmatrix} + \rho_{k+1}\begin{pmatrix}-\rho_{k+1}d_k & 0_k^T & d_k & \nu_{k,0} & \ldots & \nu_{k,n-k-2}\end{pmatrix} \\ &\quad = \begin{pmatrix}d_{k+1} & 0_{k+1}^T & \mu_{k+1,0} & \ldots & \mu_{k,n-(k+1)-1}\end{pmatrix},\end{aligned}$$

where $\mu_{k+1,0} = -\rho_{k+2}d_{k+1}$.

The Schur Algorithm

The Schur algorithm computes the scaled Cholesky factor D_nU_n of T_n with kth row given by $\begin{pmatrix}0_k^T & d_k & \nu_{k,0} & \ldots & \nu_{k,n-k-1}\end{pmatrix}$.

$$d_0 = t_0$$

$$\begin{pmatrix}\nu_{0,0} & \ldots & \nu_{0,n-1} \\ \mu_{0,0} & \ldots & \mu_{0,n-1}\end{pmatrix} = \begin{pmatrix}t_1 & \ldots & t_n \\ t_1 & \ldots & t_n\end{pmatrix}$$

$$1 \le k \le n, \quad \rho_k = -\mu_{k-1,0}/d_{k-1}, \quad d_k = d_{k-1}(1-\rho_k^2)$$

$$\begin{pmatrix}\nu_{k,0} & \ldots & \nu_{k,n-k-1} \\ \mu_{k,0} & \ldots & \mu_{k,n-k-1}\end{pmatrix} = \begin{pmatrix}1 & \rho_k \\ \rho_k & 1\end{pmatrix}\begin{pmatrix}\nu_{k-1,0} & \ldots & \nu_{k-1,n-(k-1)-2} \\ \mu_{k-1,1} & \ldots & \mu_{k-1,n-(k-1)-1}\end{pmatrix}.$$

First Class of Systolic Implementations

Assume that a linear array of $n+1$ processors, numbered 0 to n, is available for the execution of the Schur algorithm on matrix T_n of order $n+1$. A time step for the array is defined as a time interval long enough to accommodate the operations

$$\rho_k = -\mu_{k-1,0}/d_{k-1}, \qquad d_k = d_{k-1}(1-\rho_k^2)$$
$$\begin{pmatrix} \nu_{k,j} \\ \mu_{k,j} \end{pmatrix} = \begin{pmatrix} 1 & \rho_k \\ \rho_k & 1 \end{pmatrix} \begin{pmatrix} \nu_{k-1,j} \\ \mu_{k-1,j+1} \end{pmatrix}.$$

Different schedules and processor assignments for individual operations can be derived by applying appropriate linear transformations to the indices of the computed quantities: the pair $(\nu_{k,j} \quad \mu_{k,j})$ is computed in processor

$$\pi = (k \quad j) \begin{pmatrix} \pi_1 \\ \pi_2 \end{pmatrix} + \pi_3$$

at time

$$\tau = (k \quad j) \begin{pmatrix} \tau_1 \\ \tau_2 \end{pmatrix} + \tau_3.$$

The partial order of computations is preserved by observing that $\nu_{k-1,j}$ must be computed before $\nu_{k,j}$, and $\mu_{k-1,j+1}$ before $\mu_{k,j}$. That is, the time function must satisfy

$$-\tau_1 < 0, \qquad -\tau_1 + \tau_2 < 0.$$

In general, if a quantity with index (i,j) depends on a quantity with index (k,l) the latter must be available before the former can be determined, in other words [8]

$$(k-i \quad l-j) \begin{pmatrix} \tau_1 \\ \tau_2 \end{pmatrix} < 0.$$

To make sure that one processor does not have to perform two different operations at the same time the determinant of the matrix

$$\begin{pmatrix} \tau_1 & \pi_1 \\ \tau_2 & \pi_2 \end{pmatrix}$$

should be non-zero [8], $\tau_1\pi_2 - \tau_2\pi_1 \neq 0$.

The first systolic array presented in [12] is based on the following linear transformations. In step k of the Schur algorithm, $0 \le k \le n$, $(\nu_{k,j} \quad \mu_{k,j})$ is computed in processor π at time $\tau \ge 1$ where

$$\pi = (k \quad j) \begin{pmatrix} 0 \\ 1 \end{pmatrix} + 0 = j, \quad \tau = (k \quad j) \begin{pmatrix} 1 \\ 0 \end{pmatrix} + 1 = k+1, \qquad 0 \le j \le n-k-1.$$

The parameters ρ_k and d_k are assumed to be associated with index $(k, 0)$ and thus determined in processor $\pi = 0$ at time $\tau = k + 1$. Initially processor j is loaded with matrix element t_j.

The execution of the Schur algorithm requires $n + 1$ steps. In each step processor 0 computes a new ρ and broadcasts it to all other processors, so that each step produces a new row of the Cholesky factor. The components of the ν-vectors remain in their respective processors ($\nu_{k,j}$ resides in processor j for all k) while the μ-vector is shifted left by one in each step ($\mu_{k-1,j+1}$ is computed in processor $j+1$ before being sent to processor j for the computation of $\mu_{k,j}$). Note that in step k there are k idle processors, and in order to offload a row of the Cholesky factor each processor must be able to perform external I/O.

To avoid difficulties, such as synchronisation delays and long wires, associated with a global communication scheme like broadcasting a 'pipelined' array is proposed in [12, 10]. The processor function $(\pi_1 \quad \pi_2 \quad \pi_3)$ is the same as before but the time function has changed to

$$(\tau_1 \quad \tau_2 \quad \tau_3) = (2 \quad 1 \quad 1).$$

Thus, ρ_k and d_k are computed at time $\tau = 2k + 1$ in processor 0 and ρ_k is then sent to processor 1. In the next step, at $\tau = 2k+2$, $(\mu_{k,1} \quad \nu_{k,1})$ can be computed in processor 1, ρ_k can be transmitted to processor 2 and $\mu_{k,1}$ left to processor 1 so that at $\tau = 2k + 3$ the computation of ρ_{k+1} can start. Thus, successive iterations are two time steps apart. Because ρ_n and d_n are computed at $\tau = 2n+1$ the computation time for the Schur algorithm comes to $2(n + 1)$. The replacement of broadcasting by forwarding (or pipelining) of ρ from processor to processor results in communication that takes place exclusively on a nearest neighbour basis. All other features are the same as in the first array.

In order to *solve* the system $T_n x = f$ three possibilities are discussed in [12]:

1. forward elimination and backsubstitution involving the Cholesky factors U_n, D_n and U_n^T of T_n
2. computation of the Levinson vectors ψ_k using the ρ_k from the Schur algorithm, and subsequent matrix vector multiplications involving L_n, D_n, and L_n^T (the ψ_k constitute the rows of the Cholesky factor L_n of T_n^{-1})
3. explicit computation of T_n^{-1} in form of the Gohberg-Semencul formula and subsequent matrix-vector multiplications by means of FFTs (the Gohberg-Semencul formula represents the inverse of a Toeplitz matrix as a sum of products of triangular Toeplitz matrices that consist of the elements of ψ_n).

Since details for parallel implementations of the latter two methods are not given and their data and control flows are likely to be rather complex only the first method will be considered.

Forward Elimination with Cholesky Factor

Forward elimination solves the lower triangular system $(D_n U_n)^T h = f$, the elements of the solution vector h are given by $h_k = h_{k,k}$.

$$h_{0,0} = f_0/d_0$$

$$1 \le k \le n, \quad h_{k,-1} = 0$$

$$1 \le j \le k-1, \quad h_{k,j} = h_{k,j-1} + \nu_{j-1,k-j} h_{j,j}$$

$$h_{k,k} = (f_k - h_{k,k-1})/d_k.$$

In order to overlap forward elimination as much as possible with factorisation a second linear array with $n+1$ processors is employed, and it is assumed that each processor in the elimination array is physically connected to the corresponding processor in the factorisation array. Since a matrix element $\nu_{j-1,k-j}$ is computed in processor $k-j$ at time $\tau = k+j-2$ in the factorisation array it may be used at the next time step in processor $k-j$ of the forward elimination array. Hence, the forward elimination array has the processor function

$$(\pi_1 \quad \pi_2 \quad \pi_3) = (1 \quad -1 \quad 0),$$

and essentially the same time function as the factorisation array (the time functions just differ by one in their displacement τ_3):

$$(\tau_1 \quad \tau_2 \quad \tau_3) = (2 \quad 1 \quad 2).$$

Note that the elements of the h-vector are shifted one processor to the right each step. At time $\tau = 3k+2$, f_k and d_k have to be input to processor 0 of the elimination array so that $h_k = h_{k,k}$ can be computed there. Thus, forward elimination is completed after the computation of h_n at time $3n+3$.

Backsubstitution with Cholesky Factor

Backsubstitution determines the solution x, with elements $x_k = x_{k,k}$, of the upper triangular system $D_n U_n x = D_n h$.

$$x_{n,n} = d_n h_n / d_n$$

$$n-1 \geq k \geq 0, \quad x_{k,n+1} = 0$$

$$n \geq j \geq k+1, \quad x_{k,j} = x_{k,j+1} + \nu_{k,j-k-1} x_{j,j}$$

$$x_{k,k} = (d_k h_k - x_{k,k+1}) / d_k.$$

As backsubstitution can only start once forward elimination is completely finished, the forward elimination array may be re-used. Its time and processor functions are now

$$(\pi_1 \quad \pi_2 \quad \pi_3) = (-1 \quad 1 \quad 0), \quad (\tau_1 \quad \tau_2 \quad \tau_3) = (-1 \quad -1 \quad 5n+3).$$

If processor 0 has retained all h_k and d_k from the forward elimination phase then the solution element $x_k = x_{k,k}$ can be determined in processor 0 at time $\tau = 5n + 3 - 2k$. Note that solution of a Toeplitz system of order n in such a manner requires time $O(5n)$ on $2n$ processors plus $O(n^2)$ storage to store the matrix $D_n U_n$ during the forward elimination phase.

Forward Elimination by Schur Recursions

The second step, the modification of the right-hand side vector f in the system $T_n x = f$ can be improved for a systolic implementation by applying the same operations to f as were applied to T_n in the Schur algorithm: after having determined $L_n T_n = D_n U_n$ one now determines $g = L_n f$ so processors perform the same type of operations during facorisation and forward elimination, and only one type of array is needed for both phases.

To derive the computational steps for $g = L_n f$ we extend the vector f to a Toeplitz matrix F_n with f as its first column:

$$f = \begin{pmatrix} f_0 \\ f_1 \\ \vdots \\ f_n \end{pmatrix}, \qquad F_n = \begin{pmatrix} f_0 & f_1 & \cdots & \cdots & f_n \\ f_1 & f_0 & \ddots & & \vdots \\ \vdots & \ddots & \ddots & \ddots & \vdots \\ \vdots & & \ddots & f_0 & f_1 \\ f_n & \cdots & \cdots & f_1 & f_0 \end{pmatrix}.$$

From the computation of $L_n F_n$ one can derive recursions for $L_n f$ by means of the following observation. The kth element of $g = L_n f$ is

$$g_k = (\,(\,\psi_k^T \quad 1\,) \quad 0_{n-k}^T\,)\, f = (\,\psi_k^T \quad 1\,)\, f_{0:k}, \qquad 0 \le k \le n,$$

while the kth row of $L_n F_n$ is

$$\begin{aligned}(\,(\,\psi_k^T \quad 1\,) \quad 0_{n-k}^T\,)\, F_n &= (\,(\,\psi_k^T \quad 1\,) \quad 0_{n-k}^T\,) \begin{pmatrix} F_k & * \\ * & F_{n-k-1} \end{pmatrix} \\ &= (\,* \quad (\,\psi_k^T \quad 1\,)\, F_k\,)\end{aligned}$$

whose kth element is the trailing element of $(\,\psi_k^T \quad 1\,)\, F_k$ which is equal to $(\,\psi_k^T \quad 1\,)\, J f_{0:k}$. Hence, the trailing element of $(\,\psi_k^T \quad 1\,)\, J F_k$ in

$$(\,(\,\psi_k^T \quad 1\,)\, J \quad 0_{n-k}^T\,)\, F_n = (\,(\,\psi_k^T \quad 1\,)\, J F_k \quad *\,)$$

is $g_k = (\,\psi_k^T \quad 1\,)\, f_{0:k}$.

Consequently, the sought vector g is a product of f with the matrix whose rows contain the reverse Levinson vectors, and g can be computed by the following Schur-like recursions involving the upper triangular part of this matrix.

Denote elements in position $k \le i \le n$ of

$$(\,(\,\psi_k^T \quad 1\,)\, J \quad 0_{n-k}^T\,)\, F_n = (\,(\,\psi_k^T \quad 1\,)\, J \quad 0_{n-k}^T\,) \begin{pmatrix} F_k & C_k \\ C_k^T & F_k \end{pmatrix} = (\,(\,\psi_k^T \quad 1\,)\, J F_k \quad (\,\psi_k^T \quad 1\,)\, J C_k\,)$$

by

$$(\,\alpha_{k,k} \quad \ldots \quad \alpha_{k,n}\,) = (\,(\,\psi_k^T \quad 1\,)\, f_{0:k} \quad (\,\psi_k^T \quad 1\,)\, J C_k\,)$$

and elements in position $k \le i \le n$ of

$$\begin{aligned}(\,(\,\psi_k^T \quad 1\,) \quad 0_{n-k}^T\,)\, F_n &= (\,(\,\psi_k^T \quad 1\,) \quad 0_{n-k-1}^T \quad 0\,) \begin{pmatrix} F_k & D_k & J f_{n:n-k} \\ D_k^T & F_{n-k-2} & * \\ f_{n:n-k}^T J & * & f_0 \end{pmatrix} \\ &= (\,(\,\psi_k^T \quad 1\,)\, F_k \quad (\,\psi_k^T \quad 1\,)\, D_k \quad (\,\psi_k^T \quad 1\,)\, J f_{n:n-k}\,)\end{aligned}$$

by

$$(\,\beta_{k,k} \quad \ldots \quad \beta_{k,n}\,) = (\,(\,\psi_k^T \quad 1\,)\, J f_{0:k} \quad (\,\psi_k^T \quad 1\,)\, D_k \quad (\,\psi_k^T \quad 1\,)\, J f_{n:n-k}\,). \tag{8}$$

Now $\alpha_{k,k} = g_k$ is the kth element of g and the recursive computation of $\alpha_{k+1,k+1} = g_{k+1}$ from $\alpha_{k,i}$ and $\beta_{k,i}$ can be derived by means of the Levinson recursions (4) as follows

$$((\psi_{k+1}^T \quad 1) \quad 0_{n-k-1}^T)\, F_n = ((\psi_k^T \quad 1)\, J \quad 0_{n-k}^T) + \rho_{k+1}\,(0 \quad (\psi_k^T \quad 1) \quad 0_{n-k-1}^T)\, F_n.$$

Ignoring elements in positions 0 through k on both sides of the equation gives

$$(\alpha_{k+1,k+1} \quad \dots \quad \alpha_{k+1,n}) = (\alpha_{k,k+1} \quad \dots \quad \alpha_{k,n}) + \rho_{k+1}\,(\beta_{k,k} \quad \dots \quad \beta_{k,n-1})$$

since the second summand is equal to

$$(0 \quad (\psi_k^T \quad 1) \quad 0_{n-k-1}^T) \begin{pmatrix} f_0 & f_{1:k}^T & * \\ f_{1:k} & F_k & D_k \\ * & D_k^T & F_{n-k-2} \end{pmatrix} = ((\psi_k^T \quad 1)\, f_{1:k} \quad (\psi_k^T \quad 1)\, F_k \quad (\psi_k^T \quad 1)\, D_k).$$

Comparing elements in positions $k+1$ through n with (8) one notes that

$$((\psi_k^T \quad 1)\, J f_{0:k} \quad (\psi_k^T \quad 1)\, D_k) = (\beta_{k,k} \quad \dots \beta_{k,n-1}).$$

The second vector consisting of elements $\beta_{k+1,i}$, $k+1 \le i \le n$, can be updated similarly.

Forward Elimination by Schur Recursions

The Schur recursions determine $g = L_n f$ where $g_k = \alpha_{k,k}$.

$$\begin{pmatrix} \alpha_{0,0} & \dots & \alpha_{0,n} \\ \beta_{0,0} & \dots & \beta_{0,n} \end{pmatrix} = \begin{pmatrix} f_0 & \dots & f_n \\ f_0 & \dots & f_n \end{pmatrix}$$

$$1 \le k \le n, \qquad \begin{pmatrix} \alpha_{k,k} & \dots & \alpha_{k,n} \\ \beta_{k,k} & \dots & \beta_{k,n} \end{pmatrix} = \begin{pmatrix} 1 & \rho_k \\ \rho_k & 1 \end{pmatrix} \begin{pmatrix} \alpha_{k-1,k} & \dots & \alpha_{k-1,n} \\ \beta_{k-1,k-1} & \dots & \beta_{k-1,n-1} \end{pmatrix}.$$

Second Class of Systolic Implementations

Again, the same assumptions as before hold, and the array from [3, 4] is presented that performs both phases, factorisation and forward elimination, by Schur recursions.

The pipelined version of the factorisation phase is performed as before. As for forward elimination, the index structure of its equations is adapted to that of the factorisation by performing a linear transformation on each index $(k \quad j)$:

$$(k \quad j) \begin{pmatrix} 1 & -1 \\ 0 & 1 \end{pmatrix} = (k \quad j-k),$$

resulting in

$$\begin{pmatrix} \alpha_{0,0} & \dots & \alpha_{0,n} \\ \beta_{0,0} & \dots & \beta_{0,n} \end{pmatrix} = \begin{pmatrix} f_0 & \dots & f_n \\ f_0 & \dots & f_n \end{pmatrix}$$

$$1 \le k \le n, \qquad \begin{pmatrix} \alpha_{k,0} & \dots & \alpha_{k,n-k} \\ \beta_{k,0} & \dots & \beta_{k,n-k} \end{pmatrix} = \begin{pmatrix} 1 & \rho_k \\ \rho_k & 1 \end{pmatrix} \begin{pmatrix} \alpha_{k-1,1} & \dots & \alpha_{k-1,n-(k-1)} \\ \beta_{k-1,0} & \dots & \beta_{k-1,n-(k-1)+1} \end{pmatrix}.$$

If a second $(n+1)$-processor array is available for forward elimination with the same time and processor functions as those of the factorisation array, then factorisation and forward elimination can be performed simultaneously. Processor 0 of the forward elimination array is assumed to be connected to processor 0 of the factorisation array so the latter can forward the ρ_k to the former.

When pipelining is used, processor 0 of the elimination array receives ρ_k and forwards it directly to the other processors in the array so pairs $(\alpha_{k,j} \quad \beta_{k,j})$ are computed in processor j at time $2k+j+1$. Initially, processor j is loaded with the right-hand element f_j. Element $g_k = \alpha_{k,0}$ is computed in processor 0 at time $\tau = 2k+1$ and transmitted to processor 0 of the factorisation array where it is retained till the start of the backsubstitution phase. Thus, factorisation and forward elimination can be executed on $2(n+1)$ processors in $2(n+1)$ time steps if communication proceeds on a nearest neighbour basis.

In order to avoid the $O(n^2)$ storage needed to store D_nU_n till the onset of backsubstitution only its last column and the parameters ρ_k are retained from which D_nU_n can be re-generated by the Schur recursions.

The Reverse Version of the Schur Algorithm

The reverse version of Schur algorithm computes the scaled Cholesky factor D_nU_n of T_n with kth row given by $(0_k^T \quad d_k \quad \nu_{k,0} \quad \dots \quad \nu_{k,n-k-1})$ from ρ_k, $1 \le k \le n$, and the last column

$$(\nu_{0,n-1} \quad \nu_{1,n-2} \quad \dots \quad \nu_{n-1,0} \quad d_n)^T$$

of D_nU_n, whereby $\nu_{0,n-1} = t_n$.

$n-1 \ge k \ge 0,$

$$\begin{pmatrix} \nu_{k,0} & \dots & \nu_{k,n-k-2} \\ \mu_{k,1} & \dots & \mu_{k,n-k-1} \end{pmatrix} = \frac{1}{\rho_{k+1}^2 - 1} \begin{pmatrix} -1 & \rho_{k+1} \\ \rho_{k+1} & -1 \end{pmatrix} \begin{pmatrix} \nu_{k+1,0} & \dots & \nu_{k+1,n-(k+1)-1} \\ \mu_{k+1,0} & \dots & \mu_{k+1,n-(k+1)-1} \end{pmatrix}$$

$$d_k = d_{k+1}/(1-\rho_{k+1}^2), \quad \mu_{k,0} = -d_k\rho_{k+1}.$$

If processor 0 in the factorisation array has retained all ρ_k and d_n, and processor $n-k$ has received component $\nu_{k,n-k-1}$ of the last column from its left neighbour then the re-generation of $D_n U_n$ in the factorisation array can start at time $2(n+1)$. The processor and time functions are

$$(\pi_1 \quad \pi_2 \quad \pi_3) = (0 \quad 1 \quad 0), \quad (\tau_1 \quad \tau_2 \quad \tau_3) = (-2 \quad -1 \quad 4n+2),$$

so that $(\nu_{k,j} \quad \mu_{k,j+1})$ is computed in processor j at time $4n+2-2k-j$, and d_k in processor 0 at time $\tau = 4n+2-2k$. Processor 0 stores the d_k for the backsubstitution phase. Note that the components of the ν-vector stay put in a processor ($\nu_{k,j}$ resides in processor j for all k) while the components of the μ-vector are shifted one processor to the right in each step.

Suppose a third array for backsubstitution is available whose processors are connected to the corresponding processors of the factorisation array. Since $\nu_{k,j-k-1}$ is computed in processor $j-k-1$ at time $\tau = 4n+3-k-j$ it can be used at time $4n+5$ in processor $j-k$ of the backsubstitution array. With processor and time functions

$$(\pi_1 \quad \pi_2 \quad \pi_3) = (-1 \quad 1 \quad 0), \quad (\tau_1 \quad \tau_2 \quad \tau_3) = (-1 \quad -1 \quad 4n+5),$$

$x_{j,k}$ can be computed in processor $j-k$ at time $4n+5-k$. Since processor 0 has retained d_k from the previous re-generation phase and g_k is computed early enough in processor 0 of the forward elimination array (at time $\tau = 2k+1$), element $x_k = x_{k,k}$ of the solution vector can be computed in processor 0 at time $4n+5-2k$. The whole computation is completed in $4n+6$ time steps. Note that during step k of the factorisation and forward elimination phase there are $2k$ idle processors.

Therefore, $6n$ processors can compute the solution to a $n \times n$ Toeplitz system in time $O(4n)$ only relying on nearest neighbour computation, however the storage in at least one processor must be proportional to the problem size $O(n)$.

Backsubstitution

The last step is normally solved by backsubstitution $x = L_n^T D_n^{-1} g$ without making any use of the Toeplitz structure of the original system. A new approach that uses the Schur recursions also for the last step was derived in [7] and can be related to the Levinson recursions as follows.

Remember that $((\psi_{n-k}^T \quad 1) \quad 0_k^T)$, is the kth column of L_n^T, that $g_k = \alpha_{k,k}$ is the kth element of the vector g, and that d_k is the kth diagonal element of D_n, $0 \le k \le n$. With the abbreviation

$\gamma_k = g_k/d_k$, $0 \le k \le n$, one can express the solution vector as a linear combination of the columns of L_n^T:

$$x = \gamma_0 \begin{pmatrix} 1 \\ 0_n \end{pmatrix} + \gamma_1 \begin{pmatrix} \psi_1 \\ 1 \\ 0_{n-1} \end{pmatrix} + \gamma_2 \begin{pmatrix} \psi_2 \\ 1 \\ 0_{n-2} \end{pmatrix} + \ \ldots \ + \gamma_{n-1} \begin{pmatrix} \psi_{n-1} \\ 1 \\ 0 \end{pmatrix} + \gamma_n \begin{pmatrix} \psi_n \\ 1 \end{pmatrix}.$$

Define the partial sums

$$x^{(n)} = \gamma_n \begin{pmatrix} \psi_n \\ 1 \end{pmatrix}, \qquad x^{(n-k-1)} = x^{(n-k)} + \gamma_{n-k-1} \begin{pmatrix} \psi_{n-k-1} \\ 1 \\ 0_{k+1} \end{pmatrix}, \quad 0 \le k \le n-1,$$

so that $x^{(0)} = x$. It will now be shown by induction that for $1 \le k \le n$

$$x^{(n-k)} = \begin{pmatrix} 0_{n-k} \\ \epsilon_{n-k,n-k} \\ \vdots \\ \epsilon_{n-k,n} \end{pmatrix} + \begin{pmatrix} \eta_{n-k,n-k} \\ \vdots \\ \eta_{n-k,n} \\ 0_{n-k} \end{pmatrix} + \sum_{j=0}^{k} \epsilon_{n-k,n-j} \begin{pmatrix} 0_{k-j+1} \\ \psi_{n-k-1} \\ 0_{j+1} \end{pmatrix} + \sum_{j=0}^{k} \eta_{n-k,n-j} \begin{pmatrix} 0_{k-j+1} \\ J\psi_{n-k-1} \\ 0_{j+1} \end{pmatrix}. \tag{9}$$

(i) For $k = 0$ it follows from the Levinson recursion (4) that

$$\begin{aligned} x^{(n)} = \gamma_n \begin{pmatrix} \psi_n \\ 1 \end{pmatrix} &= \gamma_n \left\{ \begin{pmatrix} 0 \\ \psi_{n-1} \\ 1 \end{pmatrix} + \rho_n \begin{pmatrix} 1 \\ J\psi_{n-1} \\ 0 \end{pmatrix} \right\} \\ &= \begin{pmatrix} 0_n \\ \epsilon_{n,n} \end{pmatrix} + \begin{pmatrix} \eta_{n,n} \\ 0_n \end{pmatrix} + \epsilon_{n,n} \begin{pmatrix} 0 \\ \psi_{n-1} \\ 0 \end{pmatrix} + \eta_{n,n} \begin{pmatrix} 0 \\ J\psi_{n-1} \\ 0 \end{pmatrix}, \end{aligned}$$

where $\epsilon_{n,n} = \gamma_n = g_n/d_n$ and $\eta_{n,n} = \rho_n \gamma_n$.

(ii) Assume the statement is true for $k \ge 0$.

(iii) Using the induction hypothesis (ii) for $x^{(n-k)}$ in

$$x^{(n-k-1)} = x^{(n-k)} + \gamma_{n-k-1} \begin{pmatrix} \psi_{n-k-1} \\ 1 \\ 0_{k+1} \end{pmatrix}$$

and making use of the Levinson recursion in each of the two sums of (9) results in

$$\sum_{j=0}^{k} \epsilon_{n-k,n-j} \begin{pmatrix} 0_{k-j+1} \\ \psi_{n-k-1} \\ 0_{j+1} \end{pmatrix} = \sum_{j=0}^{k} \epsilon_{n-k,n-j} \left\{ \begin{pmatrix} 0_{k-j+1} \\ 0 \\ \psi_{n-k-2} \\ 0_{j+1} \end{pmatrix} + \rho_{n-k-1} \begin{pmatrix} 0_{k-j+1} \\ 1 \\ J\psi_{n-k-2} \\ 0_{j+1} \end{pmatrix} \right\}$$

$$= \epsilon_{n-k,n} \left\{ \begin{pmatrix} 0_{k+1} \\ 0 \\ \psi_{n-k-2} \\ 0 \end{pmatrix} + \rho_{n-k-1} \begin{pmatrix} 0_{k+1} \\ 1 \\ J\psi_{n-k-2} \\ 0 \end{pmatrix} \right\} + \sum_{j=0}^{k-1} \epsilon_{n-k,n-j-1} \left\{ \begin{pmatrix} 0_{k-j} \\ 0 \\ \psi_{n-k-2} \\ 0_{j+2} \end{pmatrix} + \rho_{n-k-1} \begin{pmatrix} 0_{k-j} \\ 1 \\ J\psi_{n-k-2} \\ 0_{j+2} \end{pmatrix} \right\}$$

for the first sum and

$$\sum_{j=0}^{k} \eta_{n-k,n-j} \begin{pmatrix} 0_{k-j+1} \\ J\psi_{n-k-1} \\ 0_{j+1} \end{pmatrix} = \sum_{j=0}^{k} \eta_{n-k,n-j} \left\{ \begin{pmatrix} 0_{k-j+1} \\ J\psi_{n-k-2} \\ 0 \\ 0_{j+1} \end{pmatrix} + \rho_{n-k-1} \begin{pmatrix} 0_{k-j+1} \\ \psi_{n-k-2} \\ 1 \\ 0_{j+1} \end{pmatrix} \right\}$$

$$= \sum_{j=0}^{k-1} \eta_{n-k,n-j} \left\{ \begin{pmatrix} 0_{k-j+1} \\ J\psi_{n-k-2} \\ 0 \\ 0_{j+1} \end{pmatrix} + \rho_{n-k-1} \begin{pmatrix} 0_{k-j+1} \\ \psi_{n-k-2} \\ 0 \\ 0_{j+1} \end{pmatrix} \right\} + \eta_{n-k,n-k} \left\{ \begin{pmatrix} 0 \\ J\psi_{n-k-2} \\ 0 \\ 0_{k+1} \end{pmatrix} + \rho_{n-k-1} \begin{pmatrix} 0 \\ \psi_{n-k-2} \\ 1 \\ 0_{k+1} \end{pmatrix} \right\}$$

for the second sum. The last term in $x^{(n-k-1)}$ expands to

$$\gamma_{n-k-1} \begin{pmatrix} \psi_{n-k-1} \\ 1 \\ 0_{k+1} \end{pmatrix} = \gamma_{n-k-1} \left\{ \begin{pmatrix} 0 \\ \psi_{n-k-2} \\ 1 \\ 0_{k+1} \end{pmatrix} + \rho_{n-k-1} \begin{pmatrix} 1 \\ J\psi_{n-k-2} \\ 0 \\ 0_{k+1} \end{pmatrix} \right\}.$$

Collecting corresponding terms gives the following expression for $x^{(n-k-1)}$

$$\begin{pmatrix} 0_{n-k} \\ \epsilon_{n-k,n-k} \\ \vdots \\ \epsilon_{n-k,n} \end{pmatrix} + \begin{pmatrix} \eta_{n-k,n-k} \\ \vdots \\ \eta_{n-k,n} \\ 0_{n-k} \end{pmatrix} + \begin{pmatrix} 0_{k+2} \\ \epsilon_{n-k,n}\psi_{n-k-2} \\ 0 \end{pmatrix} + \begin{pmatrix} 0_{k+1} \\ \rho_{n-k-1}\epsilon_{n-k,n} \\ \rho_{n-k-1}\epsilon_{n-k,n} J\psi_{n-k-2} \\ 0 \end{pmatrix}$$

$$+ \sum_{j=0}^{k-1} \begin{pmatrix} 0_{k-j} \\ 0 \\ (\epsilon_{n-k,n-j-1} + \rho_{n-k-1}\eta_{n-k,n-j})\psi_{n-k-2} \\ \rho_{n-k-1}\eta_{n-k,n-j} \\ 0_{j+1} \end{pmatrix} + \sum_{j=0}^{k-1} \begin{pmatrix} 0_{k-j} \\ \rho_{n-k-1}\epsilon_{n-k,n-j-1} \\ (\eta_{n-k,n-j-1} + \rho_{n-k-1}\epsilon_{n-k,n-j-1}) J\psi_{n-k-2} \\ 0 \\ 0_{j+1} \end{pmatrix}$$

$$+ \begin{pmatrix} 0 \\ (\gamma_{n-k-1} + \rho_{n-k-1}\eta_{n-k,n-k})\psi_{n-k-2} \\ \gamma_{n-k-1} + \rho_{n-k-1}\eta_{n-k,n-k} \\ 0_{k+1} \end{pmatrix} + \begin{pmatrix} \rho_{n-k-1}\gamma_{n-k-1} \\ (\eta_{n-k,n-k} + \rho_{n-k-1}\gamma_{n-k-1}) J\psi_{n-k-2} \\ 0_{k+2} \end{pmatrix}$$

which can be written as

$$\begin{pmatrix} 0_{n-k-1} \\ \epsilon_{n-k-1,n-k-1} \\ \vdots \\ \epsilon_{n-k-1,n} \end{pmatrix} + \begin{pmatrix} \eta_{n-k-1,n-k-1} \\ \vdots \\ \eta_{n-k-1,n} \\ 0_{n-k-1} \end{pmatrix} + \sum_{j=0}^{k+1} \begin{pmatrix} 0_{j-k+1} \\ \epsilon_{n-k-1,n-j}\psi_{n-k-2} \\ 0_{j+1} \end{pmatrix} + \sum_{j=0}^{k+1} \begin{pmatrix} 0_{j-k+1} \\ \eta_{n-k-1,n-j} J\psi_{n-k-2} \\ 0_{j+1} \end{pmatrix},$$

where

$$\begin{pmatrix} \epsilon_{n-k-1,n-k-1} & \epsilon_{n-k-1,n-k} & \cdots & \epsilon_{n-k-1,n-1} & \epsilon_{n-k-1,n} \\ \eta_{n-k-1,n-k-1} & \eta_{n-k-1,n-k} & \cdots & \eta_{n-k-1,n-1} & \eta_{n-k-1,n} \end{pmatrix}$$
$$= \begin{pmatrix} 1 & \rho_{n-k-1} \\ \rho_{n-k-1} & 1 \end{pmatrix} \begin{pmatrix} \gamma_{n-k-1} & \epsilon_{n-k,n-k} & \cdots & \epsilon_{n-k,n-1} & \epsilon_{n-k,n} \\ \eta_{n-k,n-k} & \eta_{n-k,n-k+1} & \cdots & \eta_{n-k,n} & 0 \end{pmatrix}.$$

This completes the induction.

The backsubstitution part using the Schur recursions computes the ϵ- and η-vectors and can be formulated as follows.

Backsubstitution by Schur Recursions

The Schur recursions determine the vector $x = L_n^T D_n^{-1} g$ with its kth element given by x_k.

$$\begin{pmatrix} \epsilon_{n,n} \\ \eta_{n,n} \end{pmatrix} = \begin{pmatrix} g_n/d_n \\ 0 \end{pmatrix}$$

$$1 \le k \le n, \quad \begin{pmatrix} \epsilon_{n-k,n-k} & \epsilon_{n-k,n-k+1} & \cdots & \epsilon_{n-k,n-1} & \epsilon_{n-k,n} \\ \eta_{n-k,n-k} & \eta_{n-k,n-k+1} & \cdots & \eta_{n-k,n-1} & \eta_{n-k,n} \end{pmatrix}$$
$$= \begin{pmatrix} 1 & \rho_{n-k} \\ \rho_{n-k} & 1 \end{pmatrix} \begin{pmatrix} g_{n-k}/d_{n-k} & \epsilon_{n-(k-1),n-(k-1)} & \cdots & \epsilon_{n-(k-1),n-1} & \epsilon_{n-(k-1),n} \\ \eta_{n-(k-1),n-(k-1)} & \eta_{n-(k-1),n-(k-1)+1} & \cdots & \eta_{n-(k-1),n} & 0 \end{pmatrix}$$

$$0 \le j \le n, \quad x_j = \epsilon_{0,j} + \eta_{0,j}.$$

Third Class of Systolic Implementations

The computation of all phases, factorisation, forward elimination and backsubstitution, by Schur recursions makes it possible to employ only one array for all three phases. The corresponding array in [7, 8] is the most efficient of the three types of designs presented, and can be derived as follows (the processor and time functions here differ from the ones in [7, 8] in a few small details that do not affect the asymptotic computation time).

To fit all three phases on one array it is convenient to adapt the index structure of the factorisation phase to that of forward elimination by transforming each index $(k \quad j)$ in the factorisation to

$$(k \quad j)\begin{pmatrix} 1 & 1 \\ 0 & 1 \end{pmatrix} = (k \quad k+j).$$

The transformed factorisation phase is thus expressed as:

$$\begin{aligned}
&d_0 = t_0 \\
&\begin{pmatrix} \nu_{0,0} & \dots & \nu_{0,n-1} \\ \mu_{0,0} & \dots & \mu_{0,n-1} \end{pmatrix} = \begin{pmatrix} t_1 & \dots & t_n \\ t_1 & \dots & t_n \end{pmatrix} \\
&1 \le k \le n, \quad \rho_k = -\mu_{k-1,k-1}/d_{k-1}, \quad d_k = d_{k-1}(1-\rho_k^2) \\
&\qquad \begin{pmatrix} \nu_{k,k} & \dots & \nu_{k,n-1} \\ \mu_{k,k} & \dots & \mu_{k,n-1} \end{pmatrix} = \begin{pmatrix} 1 & \rho_k \\ \rho_k & 1 \end{pmatrix} \begin{pmatrix} \nu_{k-1,k-1} & \dots & \nu_{k-1,n-2} \\ \mu_{k-1,k} & \dots & \mu_{k-1,n-1} \end{pmatrix}.
\end{aligned}$$

The time and processor functions are chosen to be

$$(\pi_1 \quad \pi_2 \quad \pi_3) = (1 \quad 0 \quad 0), \quad (\tau_1 \quad \tau_2 \quad \tau_3) = (1 \quad 1 \quad 2).$$

Thus, all matrix elements are input to the same processor: t_k is input to processor 0 at time $\tau = k+1$; and $(\nu_{k,j} \quad \mu_{k,j})$ are determined in processor k at time $\tau = k+j+2$. The values of ρ_k and d_k are computed along with $\nu_{k,k}$, in processor k at time $2k+2$, and remain in that processor throughout factorisation and forward elimination. Notice that the components of the ν-vector stay put in the processor while the components of the μ-vector are shifted one processor to the left. The last quantities ρ_n and d_n are computed in processor n at time $2(n+1)$, so the computation of the factorisation requires $2n+3$ steps.

Since the factorisation has the same structure as the forward elimination phase, and the forward elimination phase involves the ρ_k which are now computed in different processors the two phases may be overlapped, thereby eliminating the processor idle time of the previous designs. Observe that the last matrix element t_n is input to processor 0 at $\tau = n+2$ so the first element of the right-hand side vector f_0 can be input to processor 0 at time $n+3$. In general, all right-hand side elements are input to the same processor as the matrix elements: f_j is input to processor 0 at time $n+j+3$, and time and processor functions (except for the time displacement τ_3) are the same as before:

$$(\pi_1 \quad \pi_2 \quad \pi_3) = (1 \quad 0 \quad 0), \quad (\tau_1 \quad \tau_2 \quad \tau_3) = (1 \quad 1 \quad n+3).$$

The pair $(\alpha_{k,j} \quad \beta_{k,j})$ is determined in processor k at time $\tau = k+j+n+3$, and the components of both α- and β-vectors experience a shift to the left neighbouring processor after their computation. Element $g_k = \alpha_{k,k}$ is computed in processor k at time $2k+n+3$, and forward elimination is completed at time $3n+4$.

To keep communication on a nearest neighbour basis, the linear array is folded together so that processors k and $n-k$ are situated across from each other. After completion of the forward elimination phase processors k and $n-k$ can then exchange their values of ρ, d and g so that processor k ends up with ρ_{n-k}, d_{n-k} and g_{n-k}. For simplicity each index $(k \quad j)$ of backsubstitution is transformed to

$$(k \quad j)\begin{pmatrix} -1 & 0 \\ 0 & 1 \end{pmatrix} + (n \quad 0) = (n-k \quad j),$$

resulting in

$$\begin{pmatrix} \epsilon_{0,n} \\ \eta_{0,n} \end{pmatrix} = \begin{pmatrix} g_n/d_n \\ 0 \end{pmatrix}$$

$$1 \le k \le n, \quad \begin{pmatrix} \epsilon_{k,n-k} & \epsilon_{k,n-k+1} & \cdots & \epsilon_{k,n-1} & \epsilon_{k,n} \\ \eta_{k,n-k} & \eta_{k,n-k+1} & \cdots & \eta_{k,n-1} & \eta_{k,n} \end{pmatrix}$$

$$= \begin{pmatrix} 1 & \rho_{n-k} \\ \rho_{n-k} & 1 \end{pmatrix} \begin{pmatrix} g_{n-k}/d_{n-k} & \epsilon_{k-1,n-(k-1)} & \cdots & \epsilon_{k-1,n-1} & \epsilon_{k-1,n} \\ \eta_{k-1,n-(k-1)} & \eta_{k-1,n-(k-1)+1} & \cdots & \eta_{k-1,n} & 0 \end{pmatrix}$$

$$0 \le j \le n, \quad x_j = \epsilon_{n,j} + \eta_{n,j}.$$

With processor and time functions

$$(\pi_1 \quad \pi_2 \quad \pi_3) = (1 \quad 0 \quad 0), \quad (\tau_1 \quad \tau_2 \quad \tau_3) = (2 \quad 1 \quad 2n+4)$$

the pair $(\epsilon_{k,j} \quad \eta_{k,j})$ is computed on processor k at time $\tau = 2k+j+2n+4$. In particular, component $x_k = \epsilon_{n,k} + \eta_{n,k}$ of the solution vector is computed in processor n at time $\tau = 4(n+1)+k$. Hence backsubstitution is completed at time $5n+6$.

With the above scheme, a Toeplitz system of order n can be solved in time $O(5n)$ on n processors with nearest neighbour commuincation. Each processor requires only a constant amount of storage. External input takes place on the first processor and external output on the last. As shown in [7, 8] the solution processes for several different problems with different right-hand sides can be overlapped and the solution to a new problem can be obtained every n steps. Furthermore, as shown in [8], the above array belongs to the class of n-processor arrays that solve Toeplitz systems faster than any other array with linear processor and time function, and I/O restricted to the boundary processors.

Acknowledgements

The work presented in this paper was supported by the Office of Naval Research under contracts N000014-86-K-0310 and N00014-85-K-0461, and by the Army Research Office under contract DAAL03-86-K-0158.

References

[1] Bareiss, E.H., *Numerical Solution of Linear Equations with Toeplitz and Vector Toeplitz Matrices,* Numer. Math., 13 (1969), pp. 404–24.

[2] Brent, R.P., Gustavson, F.G. and Yun, D.Y.Y., *Fast Solution of Toeplitz Systems of Equations and Computation of Pade Approximants,* J. Algorithms, 1 (1980), pp. 159–95.

[3] Brent, R.P., Kung, H.T. and Luk, F.T., Some Linear-Time Algorithms for Systolic Arrays, *Proc. IFIP 9th World Computer Congress,* North Holland, Amsterdam, 1983, pp. 865–76.

[4] Brent, R.P. and Luk, F.T., *A Systolic Array for the Linear-Time Solution of Toeplitz Systems of Equations,* J. VLSI and Computer Systems, 1 (1983), pp. 1–22.

[5] Bunch, J.R., *Stability of Methods for Solving Toeplitz Systems of Equations,* SIAM J. Sci. Stat. Comput., 6 (1985), pp. 349–64.

[6] Delosme, J.-M., *Algorithms for Finite Shift-Rank Processes,* Ph.D. Thesis, Dept of Electrical Engineering, Stanford University, 1982.

[7] Delosme, J.-M. and Ipsen, I.C.F., *Parallel Solution of Symmetric Positive Definite Systems with Hyperbolic Rotations,* Linear Algebra and its Applications, 77 (1986), pp. 75–111.

[8] ———, Efficient Systolic Arrays for the Solution of Toeplitz Systems : An Illustration of a Methodology for the Construction of Systolic Architectures in VLSI, *Systolic Arrays,* Adam Hilger, 1987, pp. 37–46.

[9] Delosme, J.-M. and Morf, M., Normalized Doubling Algorithms for Finite Shift-Rank Processes, *Proc. 20th IEEE Conference on Decision and Control,* 1981, pp. 246–8.

[10] Grenander, U. and Szego, G., *Toeplitz Forms and their Applications,* University of California Press, 1958.

[11] Iohvidov, I.S., *Hankel and Toeplitz Matrices and Forms,* Birkhauser, 1982.

[12] Kung, S.-Y. and Hu, Y.H., *A Highly Concurrent Algorithm and Pipelined Architecture for Solving Toeplitz Systems,* IEEE Trans. Acoustics, Speech, and Signal Processing, ASSP-31 (1983), pp. 66–76.

[13] Kung, H.T. and Leiserson, C.E., Systolic Arrays (for VLSI), *Sparse Matrix Proceedings,* SIAM, Philadelphia, PA, 1978, pp. 256–82.

[14] Levinson, N., *The Wiener RMS (Root-Mean-Square) Error Criterion in Filter Design and Prediction,* J. Math. Phys., 25 (1947), pp. 261–78.

[15] Morf, M., *Fast Algorithms for Multivariable Systems,* Ph.D. Thesis, Dept of Electrical Engineering, Stanford University, 1974.

[16] Nash, J.G., Hansen, S. and Nudd, G.R., VLSI Processor Array for Matrix Manipulation, *CMU Conference on VLSI Systems and Computations,* Computer Science Press, 1981, pp. 367–73.

[17] Roebuck, P.A. and Barnett, S., *A Survey of Toeplitz and Related Matrices,* Int. J. Systems Sci., 9 (1978), pp. 921–34.

[18] Schur, I., *Ueber Potenzreihen die im Innern des Einheitskreises Beschraenkt Sind,* J. Reine Angewandte Mathematik, 147 (1917), pp. 205–32.

[19] Trench, W.F., *An Algorithm for The Inversion of Finite Toeplitz Matrices,* J. Soc. Indust. Appl. Math., 12 (1964), pp. 515–22.

ENSEMBLE ARCHITECTURES AND THEIR ALGORITHMS: AN OVERVIEW

S. Lennart Johnsson
Departments of Computer Science
and Electrical Engineering
Yale University

Abstract

During recent years the number of commercially available parallel computer architectures have increased dramatically. The number of processors in these systems vary from a few up to $64k$ processors for the Connection Machine. In this paper we discuss some of the technology issues that are the underlying driving force, and focus on a particular class of parallel computer architectures often called Ensemble Architectures. They are interesting candidates for future high performance computing systems. The ensemble configurations discussed here are linear arrays, 2-dimensional arrays, binary trees, shuffle-exchange networks, Boolean cubes and cube connected cycles. We discuss a few algorithms for arbitrary data permutations, and some particular data permutation and distribution algorithms used in standard matrix computations. Special attention is given to data routing. Distributed routing algorithms in which elements with distinct origin and distinct destinations do not traverse the same communications link make possible a maximum degree of pipelined communications. The linear algebra computations discussed are: matrix transposition, matrix multiplication, dense and general banded systems solvers, linear recurrence solvers, tridiagonal system solvers, fast Poisson solvers, and very briefly, iterative methods.

1 Introduction

Advances in device technology have thus far been the primary factors in the evolution of high performance computing. Switching times have decreased from 1 μs for vacuum tubes to around $0.1 - 0.05 ns$ for MOS technologies, and an order of magnitude less for bipolar technologies. Clock rates have increased from below 1 MHz to 10 – 40 MHz in MOS technologies, and 250 MHz for the CRAY-2 (bipolar). While switching speeds have increased by four to five orders of magnitude (six for the CRAY-2), clock rates have increased by only two to three orders of magnitude. The instruction issue rate has increased similarly to the clock rates, but the amount of work carried out per instruction varies significantly. The rate at which floating-point operations can be performed has increased from 0.1 – 1 kflops (floating-point operations per second) to 10 – 20 Mflops for MOS technologies, and 250 Mflops per floating-point unit in a CRAY-2.

Of the improvement of five to six orders of magnitude in floating-point capability at most three orders of magnitude are attributed to technological and low level design improvements. Dedicated hardware for floating-point operations, extensive use of pipelining, and enhanced algorithms contribute the remainder. In current high performance systems a floating-point operation can be initiated every clock cycle. Silicon technologies are expected to offer about one order of magnitude increased switching speed before fundamental limits are reached. Other technologies, such as gallium arsenide, potentially offer a further five to ten fold increase in switching speeds.

Further dramatic increases in performance must derive from architectural innovations that exploit concurrency in computations. The critical path in most designs is determined by wire delays rather than gate delays. Comparing common microprocessor designs with a highly optimized design such as the CRAY one cannot expect to reduce the number of switching delays per clock cycle by more than a factor of 5. Reducing the clock cycle through architectural means, i.e., decreasing the number of switching delays per clock cycle, gets increasingly difficult because wire delays are ever more significant as feature sizes are reduced. Wire delays do not scale well if the geometric aspect ratios and the electric field in the gate region remain constant. While switching delays may be reduced in proportion to the scaling, wire delays may either decrease slowly, remain constant, or even increase depending on what factors (capacitive or resistive) govern the delay.

Replication of locally interconnected parts offers the highest promise for high performance architectures.

	Chip $25mm^2$ $100M\lambda^2$	Chip $50mm^2$ $200M\lambda^2$	4" Wafer (50%) $166G\lambda^2$	Clock
Dynamic RAM	1 Mbit	2 Mbit	160 Mbit	
Static RAM	256 Kbit	512 Kbit	40 Mbit	
16-bit proc.	40	80	6400	54 MHz
32-bit proc.	8	16	1280	36 MHz

Table 1: Chip and wafer level integration, 1μ feature size

	Chip $25mm^2$ $1600M\lambda^2$	Chip $50mm^2$ $3200M\lambda^2$	4" Wafer (50%) $256G\lambda^2$	Clock
Dynamic RAM	16 Mbit	32 Mbit	2660 Mbit	
Static RAM	4 Mbit	8 Mbit	640 Mbit	
16-bit proc.	640	1280	102400	216 MHz
32-bit proc.	128	256	20480	144 MHz

Table 2: Chip and wafer level integration, 0.25μ feature size

Intermingling storage and processing elements reduces the average area per processing element, and increases the clock rate in synchronous (non-pipelined) designs. The increased ratio of processing capability per unit of storage, and the increased clock rate, both contribute to increasing the maximum size of the state that can change in a single clock cycle, i.e., the rate of computation.

As feature sizes are reduced the amount of storage and logic that fit on a single chip or wafer become impressive. Tables 1 and 2 contain predictions for 1μ and 0.25μ feature sizes, respectively. The basis for the storage predictions are that a 1-bit dynamic RAM cell requires an area of $100\lambda^2$ and a 1-bit static RAM cell requires an area of $400\lambda^2$. Processor estimates are based on the Caltech Mosaic 16-bit processor, which is about $2.5M\lambda^2$, excluding the pad frame,[93], and the RISC and MIPS 32-bit processors, which are about $12M\lambda^2$, excluding pad frame [34,35,71].

At the 0.25μ feature size level a large number of processors fit on a single die. The estimates in the tables are highly simplistic and ignore the area required for interprocessor communication as well as processor-to-storage communication. With communication in the form of a one- or two-dimensional array of processors with local storage the wiring area should not substantially alter the estimates. With other interprocessor and/or processor-to-storage communication networks, substantial area may be required for wiring. Assuming bit-serial communication and simple switching elements for an Ω-network, or a Boolean cube, and the design of [73] as a base case we obtain the figures in Table 3. The predictions for the Cosmic Cube are based on an estimated processor area of $140M\lambda^2$ [119], and measured performance on existing hardware.

In any fabrication process it is expected that some of the processing cells will be defective. In a two-dimensional array of cells on a wafer in which bad cells are arbitrarily distributed, it may still be possible to use the wafer by configuring wires around the defective cells, for example, by laser-restructuring techniques. It is desirable to design wafers so that live cells can be configured in the desired pattern by "threading around" the dead cells. Leiserson and Leighton [86] and independently Greene and El Gamal[32] have investigated the problem of configuring one- and two-dimensional arrays of processors on a faulty wafer. They show that if the faulty nodes are randomly, and independently distributed, then the live nodes may be connected into a smaller two-dimensional mesh with expected maximum edge length $O(logN)$ and a channel of width $O(1)$, where N is the number of cells in the array. Simple regular structures are not costly to assemble under the presence of faults.

32-bit RISC II processors	160 Mbyte 10240 processors 370 GIPS
16-bit MOSAIC processors	160 Mbyte 51200 processors 55 GFLOPS 32-bit add 12 GFLOPS 32-bit mult
Cosmic Cube nodes ($140M\lambda^2$) (Intel 8086, 8087, 128kbyte)	256 Mbyte 2000 processors 2 GFLOPS ("measured")

Table 3: Wafer level integration, same area for processors and storage

1.1 Ensemble Architectures

Ensemble architectures [118] represent a low cost alternative to future high performance systems. High nominal performance at low cost is obtained by composing systems out of a large number of parts mass produced in state-of-the-art technology. The storage may be entirely distributed among the processors, or part of the storage may be subdivided into storage modules, each of which forms a node in a network. In the generic architecture some of the nodes in the network represent processors with storage, others storage alone. The network topology in an ensemble architecture is sparse and regular. Control and data are distributed. The notion of ensemble architectures is not new. PEPE [69,48] is an early example of an ensemble architecture of medium granularity, and cellular automata can be viewed as ensemble architectures of fine granularity.

High performance requires a high rate of operand consumption and generation. With only a few operations per operand high storage bandwidth is required. High storage bandwidth is achieved in ensemble architectures through a highly partitioned storage. Associative memory is one extreme instance of this philosophy in that each storage cell is equipped with some processing logic. Systolic architectures and the Connection Machine [38,37] are close to this extreme in that there are only a few registers, or a limited amount of storage, per processing node. The storage bandwidth of the Connection Machine model CM-I is 32 $Gbytes/sec$ at 4 MHz. Model CM-II offers a peak bandwidth in excess of 50 $Gbytes/sec$. Intermediate levels of storage bandwidth are obtained by a larger granularity of computations, as in the Cosmic Cube. Partitioning of storage to yield storage bandwidth compatible to processor capacity is used in high performance architectures such as the CRAY and CYBER series of computers, and was used already in early computer designs [76]. The partitioning of the storage is much less than in, for instance, the Connection Machine, and so is the storage bandwidth, which is 4 $Gbytes/sec$ for the CRAY-2.

There are many considerations in choosing interconnection networks, processor designs, and programming models for ensemble architectures. Manufacturability and scalability with respect to performance and reduced feature sizes of the technology are assured by interconnecting the processing elements sparsely and regularly. The interconnection network chosen represents a tradeoff between communication bandwidth, fault-tolerance, and design and manufacturing considerations. The global architecture (SIMD or MIMD), the granularity of nodes and their architectural features must be chosen so that high real performance can be achieved by minimizing communication and computation time.

1.1.1 Interconnection Networks

The choice of interconnection network determines the rate at which processors can communicate. While high bandwidth is desirable from the point of view of an algorithm designer, it may be undesirable, and in fact infeasible, from the point of view of the hardware designer. High degree of interconnect adversely affects scalability, area and volume requirements, and clock rate. One of the most pressing problems in assembling large systems across many chips is caused by severe pin restrictions — while the number of components per chip is expected to grow by up to two orders of magnitude, the chip sizes are not expected to increase much and with pin sizes limited by mechanical considerations the bandwidth at the boundary of a chip will only increase if the rate at which off-chip wires are driven is increased. Unless many communication channels are multiplexed per pin, thereby making the system clock longer, high fanout systems simply cannot be built.

There is an intimate relationship between processor design and the choice of interconnection network. As the capacity of local storage grows, so does the interprocessor distance, and the distance to the furthest location of local storage. With increased local storage it becomes necessary to structure storage [94] to minimize access delay. In the capacitive model for wire delays the access time can be reduced to order $logM$ and in the resistive model to order $\sqrt{M}$ for a storage of size M. Assume for the moment that the processors with local storage are interconnected as a one- or two-dimensional array. Then, the interprocessor distance grows as $\sqrt{M}$ and the minimum time to drive the interprocessor connection increases in proportion to $logM$ for the capacitive model and $\sqrt{M}$ for the resistive model. There is no qualitative difference in the relative growths in local access time and interprocessor communication time. Small local storage allows for short wires between processors, and potentially high clock rates. For one- or two-dimensional arrays it is desirable to keep the local storage small. Local storage can increase performance, if arithmetic is parallel and communication serial, because with local storage several such references are typically made for each remote reference.

Many interconnection networks with a small diameter, such as the shuffle-exchange, butterfly, cube connected cycles and Boolean cube require long interconnections, when layed out in a two- or three-dimensional space. Interprocessor communication will be slower than references to local storage, even if the whole network were to fit on a single wafer in submicron technology. It is important that such systems are self-timed in order for computations to make use of the higher bandwidth between a processor and its local storage, than the bandwidth for remote references. Note that the access time for different remote references may differ. The fact that some of the networks with a high wiring area do not make effective use of the area (silicon), may indeed cause networks like a mesh to offer a higher total bandwidth than, for instance, a Boolean cube [106].

Network costs

Configuring processors as linear arrays and complete binary trees requires a total number of interconnections equal to the number of processors. Both configurations scale in an excellent way. With several processors on a single chip, the required bandwidth at the chip boundary only grows at the rate of the clock frequency, regardless of the number of processors per chip and the size of the machine being built [87]. The tree has the advantage over the linear array that its diameter (the distance between the processors that are furthest apart) is $2(log_2N - 1)$ compared to N (or $\frac{1}{2}N$ for an array with end-around connections). The diameter of the network topology defines a lower bound for the speed of computation [29]. Global communication can be accomplished faster in a complete binary tree than in a linear array. The required area for the complete binary tree is of order $O(N)$ [94], if the nodes can be placed arbitrarily in the plane, in which case the maximum wire length is $\frac{1}{logN}\sqrt{N}$ [100,9]. Placing all the leaf nodes of the complete binary tree along the boundary yields an area requirement of order $O(Nlog_2N)$ [11] and a maximum wire length of order $O(\frac{1}{logN}N)$.

Linear arrays and complete binary trees have small bandwidth and present communication bottlenecks for many important computations. The two-dimensional mesh and mesh of trees [85] offer higher bandwidth and are preferable for many matrix computations. The first two networks can be realized in small area on a wafer ($O(N)$ for the N node mesh and $O(N\log^2 N)$ for the N node mesh of trees) with wire lengths $O(1)$ for the mesh and $O(\frac{1}{\log\log N}\sqrt{N}\log N)$ for the mesh of trees. The advantage of the mesh of trees over the mesh is its logarithmic diameter ($2\log N$ compared with $2\sqrt{N}$ for the mesh).

More sophisticated networks, such as the shuffle-exchange, Ω-networks, cube connected cycles [102], and Boolean n-cube have also been proposed because of their ability to efficiently emulate other important networks, or for high total bandwidths. They have been extensively studied in the literature. All these networks require layout area almost quadratic in the number of nodes, and wire lengths that grow almost linearly with the number of nodes. Correspondingly, the cost per communication is extremely high and the clock rates are decreased. Currently, a 64 input one bit wide Ω-network with simple switching elements, or

Configuration	Nodes	Diam	Fan-out	Edges
Linear Array	2^k	2^{k-1}	2	$2^k - 1$
2-d mesh	2^k	$2(2^{k/2} - 1)$	4	$2(2^k - 2^{k/2})$
2-d mesh of trees	$3 \cdot 2^k - 2 \cdot 2^{\frac{k}{2}}$	$2k$	6	$5 \cdot 2^k - 4 \cdot 2^{k/2})$
Tree of meshes	$(k+1)2^k$	$8 \cdot 2^{\frac{k}{2}}$	4	$2(2^k(4k-1) + \sqrt{2^{k+1}})$
Binary tree	$2^k - 1$	$2(k-1)$	3(1)	$2^k - 2$
Boolean cube	2^k	k	k	$k \cdot 2^{k-1}$
CCC	$k2^k$	$2k - 1$	3	$3k \cdot 2^{k-1}$
Shuffle-exchange	2^k	$2k - 1$	≤ 3	$\approx 1.5 \cdot 2^k$

Table 4: Topological properties of some common networks

Configuration	Nodes	Edge len.	Area	Pin Count
Linear Array	2^k	$O(1)$	$O(2^k)$	2
2-d mesh	2^k	$O(1)$	$O(2^k)$	$4\sqrt{M}$
2-d mesh of trees	$K = 3 \cdot 2^k - 2 \cdot 2^{\frac{k}{2}}$	$O(\sqrt{K} logK/loglogK)$	$O(Klog^2 K)$	$\approx \frac{6}{\sqrt{3}}\sqrt{M}$
Tree of meshes	$K = (k+1)2^k$	$O(k + logk)$	$O(KlogK)$	$2^{\lceil \frac{m}{2} \rceil}$
Binary tree	$2^k - 1$	$O(2^{k/2}/k)$	$O(2^k) - O(2^k \cdot k)$	4
Boolean cube	2^k	$O(2^{\frac{k}{2}})$	$O(2^{2k})$	$M(k - logM)$
CCC	$k2^k$	$O(2^{\frac{k}{2}})$	$O(k^2 2^{2k})$	$M - \frac{M}{k} log_2 \frac{M}{k}$
Shuffle-exchange	2^k	$O(2^k/k)$	$O(2^{2k}/k^2)$	

Table 5: Layout properties of some common networks

2 32-input Batcher sorting networks, fits on a single chip [73].

Large systems must be partitioned across many chips and boards. Not all the networks mentioned above are easily partitioned under the existing or predicted pin constraints. Tables 4 and 5 summarize the above discussion. The number of off-chip channels are stated in terms of the number of processors per chip, M.

Nodes in a network may either be switching elements or complete processors. The two alternatives yield architectures with slightly different properties. Multistage shuffle-exchange networks, Ω-networks, banyan networks, and butterfly networks are all closely related interconnection networks in which internal nodes typically are switching elements, possibly with a queueing capacity. These networks are used in architectures such as the HEP [10], the Ultracomputer [117,31], the RP3 [101], TRAC [120], CEDAR [78], and the BBN Butterfly [20]. In some designs processors with local storage are at both ends of the network, in others the processors have a negligable amount of storage, and storage units and processors are at opposite sides of the network, and at others again processors with a measurable amount of storage are at one side of the network and a "shared" storage at the other side. The CHiP architecture [81] represent an "intermediate" form of architecture in that it uses a switch network for communication between processors with local storage, but the switch network is novel and any path between a restricted pair of processors goes through a fixed number of switches, like 2 or 4. The switch network yields a capability to reconfigure the ensemble into a large variety of common configurations. Recently, Leiserson has proposed the Fat-tree network [88] as a universal hardware-efficient network that also uses switches, but with rebroadcasting instead of queues at switching nodes.

Network capabilities

An attractive feature of a network is the ease and efficiency with which it can simulate other networks. If the simulation does not require much overhead, and if the processors of the network being simulated can be mapped automatically onto the existing network, then any program written for the first can be automatically compiled to run efficiently on the second. The problem of finding a communication efficient algorithm for a specific network can then be formulated as a problem of embedding one graph, the *guest* graph corresponding to the communication needs of the algorithm in the graph describing the network, the *host* graph. Typically edges in the guest graph are mapped onto paths in the host, and the host may have a larger set of nodes than the guest. The *dilation* of an edge is equal to the length of the path it is mapped to in the host graph, and the *expansion* is the ratio of the number of nodes in the host and guest graphs. The dilation of edges can cause a corresponding decrease in throughput (the time between successive computations of a given kind), or just an increase in latency (additional time for completion of the first computation in a set).

For elementary algorithms common data structures and communication patterns are one- to four-dimensional meshes, complete or arbitrary binary trees, the FFT butterfly network, and the data manipulator network. As may be expected, hosts with a low connectivity and small bisection width, such as one-dimensional arrays and trees, are not efficient universal networks. Two-dimensional arrays of processors are better hosts. Thompson gives an embedding of the FFT butterfly network in a two-dimensional mesh such that $logN$ nodes of the butterfly are mapped to one node in the mesh. The maximum edge dilation is $\sqrt{N}$, and up to $logN$ butterfly edges are mapped onto one edge of the mesh.

The Boolean cube is an exceptionally versatile host graph. Multi-dimensional arrays can be embedded with dilation 1 and expansion 1 in the Boolean cube, if the number of grid points in each dimension of the array is a power of 2. It is also well known that the FFT butterfly network can be embedded in the Boolean cube with $logN$ butterfly nodes per cube node, such that the dilation is 1 and there is a one-to-one correspondence between edges in the FFT network and the cube. A static embedding allows a normal [129] algorithm (one that proceeds from input to output without reversing direction) to execute in the same number of steps on the cube as on the FFT butterfly network, but the throughput of the cube is lower by a factor of $logN$, since a cube node simulates $logN$ FFT network nodes. Similarly, a dynamic embedding of the FFT-butterfly network in a shuffle-exchange network yields a slowdown by a factor of 2 for normal algorithms. The throughput is degraded by a factor of $2logN$. Similar results hold for bitonic sorting networks, being recursively composed FFT networks. Recently, a number of results has been obtained on the embedding of complete binary trees, multiple complete binary trees, multiple binomial trees, balanced spanning trees, and arbitrary trees for the cube [132,8,7,50,44]. It follows from [8] that the Boolean cube can efficiently simulate the mesh of trees network. By construction of the tree of meshes network it is not difficult to see that the Boolean cube can embed this network with dilation 1. The universality of the Boolean cube follows from the fast implementation of sorting algorithms, and the randomized routing schemes of Valiant [131].

Applications rarely consist of a single type of computation. Each component of the set of "elementary" computations defining an application may have different ideal data structures. Indeed, it may even be the case that different data structures are ideal for different phases of an "elementary" algorithm, since the communication pattern may not be uniform throughout the execution of the algorithm [58]. The need for data reallocations in order to minimize the complexity of an algorithm decreases with the ability of the newtwork to efficiently support different access patterns to a data structure. The Boolean cube can emulate with low overhead many of the prototypical graphs that either represent particular data structures or data dependency graphs for standard algorithms, or networks that have been proposed as VLSI computing structures. A static embedding of a data structure can support many types of access schemes without communication penalty, reducing the need for data reallocations.

For arbitrary computations one measure of the utility of a network is its total bandwidth, which is proportional to the total number of edges. Hence, the bandwidth of a pipelined Ω-network, (for example, the NYU Ultracomputer [31]), is the same as the bandwidth of a Boolean cube. However, there is a latency in the switch that grows logarithmically with the number of processing elements. In the Boolean cube the interprocessor communication time is nonuniform. The minimum interprocessor communication time amounts to one routing. The maximum number of routing steps is log_2N. Interprocessor communication in an Ω-network includes $2log_2N$ links. If the communication width corresponds to a word, then this latency may be a significant fraction of the communication time. With bit-serial communication the difference may be negligable, and entirely dependent upon various implementation decisions. Moreover, whether the nominal difference results in a real performance difference depends on the particular data dependences of the computation, and the mapping of the computations on to the architecture.

2 Ensemble Architecture Algorithms

An ensemble architecture of extreme concurrency is similar to systolic architectures. However, in ensemble architectures data management is an even more predominant factor. In most, but not all, systolic architectures most of the data (input and output) are stored outside the array, and the management of such data is generally ignored. In algorithms for ensemble architectures it is generally assumed that initial data, as well as the results, are stored within the ensemble. Furthermore, the number of nodes in the ensemble is, in general, insufficient to match characteristic parameters of the problem, but the amount of storage per node is significant. The granularity of computations in ensemble architectures is often larger than in systolic architectures.

Time-space trade-offs are at the core of mapping algorithms onto ensemble architectures. Data and control structures, synchronization and communication are, in general, considerably more complex in ensemble architectures than in systolic designs. The time-space trade-off in ensemble architectures is often made in favor of minimizing data movement. Systolic designs are of fine grain and designs are often such that the communication time is comparable to the time for logic or arithmetic operations. Most ensemble architectures are of a coarser grain and interprocessor communication is typically slower than the execution of arithmetic and logic operations.

Algorithms are devised both in an ad hoc manner and systematically. The first approach may lead to entirely new, efficient, algorithms. The second approach can be supported by algorithm design tools, and provides the necessary insight to develop compilation techniques that transform abstract representations of algorithms into efficient code for a variety of architectures. In the systematic approach, which is followed in the description of sample algorithms below, a *computation graph* defining the partial ordering of computations is created from the definition of the computation in a suitable notation. Then, this computation graph is mapped on to the ensemble. The computation graph has a *level* or *stage* for each sequential step of the algorithm. Nodes of the graph represent computations, and the computations represented by the nodes at a given level can all be performed concurrently. Arcs between nodes represent data transfer, which with nodes of the computation graph mapped into different processors represent communication. In a sufficiently large ensemble the mapping of the computation graph can be made such that all nodes of a given level are assigned to distinct processors. The situation in this case is similar to what is typical for systolic designs [68], [65], [17], [91], [97], [96], [103], [89], [19], [22]. If there are fewer processors than the maximum number of nodes at any one level of the computation graph, then different nodes have to be identified with the same processor. Two such schemes are *cyclic* and *consecutive* identification [50] defined precisely later.

In the identification of multiple nodes of the computation graph with processors in a specific ensemble architecture several performance related issues arise that do not occur in designs of the systolic type. For instance, in such a design it is often sufficient for maximum utilization of resources that no two elements compete for the same communications link at the same stage in the execution of the algorithm. But, in an ensemble architecture it may be required that for maximum performance communications during different stages of the algorithm do not compete for the same communication link. The ability to establish different communication paths with a minimum number of shared edges becomes important. The size of the problem relative to the size of the ensemble also affects the optimum embedding in other ways due to restricted communication.

With larger granularity of computations operations are no longer occurring concurrently to the extent disclosed by the computation graph. A parallel algorithm that minimizes the time for arithmetic operations on an unbounded number of processors may have a higher total operations count than an algorithm minimizing the number of arithmetic operations. Bitonic sort and odd-even cyclic reduction are examples thereof. In order to minimize the required solution time on an ensemble of finite size a combination of algorithms may be needed. In some instances such combinations can be obtained through algorithm transformations. Which algorithm minimizes the execution time may also depend upon the number of problems to be solved in that some algorithms are more amenable to pipelining than others.

We will describe some basic ensemble architecture algorithms for computational linear algebra and sorting, and focus on the issues raised above. The ensemble architecture topologies used as model architectures are linear arrays, 2-dimensional meshes, binary trees, shuffle-exchange, Boolean cube, and cube connected cycles networks. The algorithms have communication topologies in the form of *one- or multi-dimensional meshes, butterfly networks, data manipulator networks, and spanning trees.* We first describe the embeddings of these graphs in graphs representing the topology of the ensemble architectures.

2.1 Graph Embeddings

The computation graph defines a partial ordering of the computations. Constraints on the realization of the computation graph are imposed by the ensemble architecture, and are incorporated in the mapping process. The computations corresponding to the nodes at a given level (order) have to be spread over time if there is an insufficient number of nodes in the ensemble, or if the communication implied by directed edges may require more than one communication step. This situation occurs if the operands at the source and sink of the edge are located at nodes at a distance greater than 1 in the ensemble. In the interest of conserving storage, nodes of the computation graph are sometimes identified with a given storage location. Such a strategy results in a variety of access schemes for the same data structure. If the storage has a latency, then the latency may determine the rate of execution during some part of the algorithm as for FFT and odd-even cyclic reduction [15] on vector architectures. The bank conflict problem in vector processors is well known, and architectural solutions [13], [82], [84], [105], as well as solutions at the application program level for particular algorithms [72], [47] have been proposed.

The situation is the same within a node in an ensemble architecture, but in addition the time of accessing storage is not uniform. In a simplified model the storage of nodes with which a given node has direct connections can be accessed in 1 unit of time, the storage of the neighboring nodes of the immediate neighbors in 2 units of time, etc. The larger the number of neighbors the larger the number of different access schemes that can be supported by a fixed data structure at a given number of communication actions. Reconfigurability through switchable interconnections gives the ensemble the same property.

2.1.1 Complete Binary Tree Hosts

We first consider guest graphs in the form of one- and multi-dimensional arrays. Rosenberg and Snyder [109] and Sekanina [121], have given a procedure for a *proximity preserving* embedding of a $2^n - 1$ node loop in a $2^n - 1$ node binary tree. Let $d_L(i,j)$ be the distance between nodes i and j in the loop, and $\phi(i)$ and $\phi(j)$ be the indices of the tree nodes to which nodes i and j are mapped. Then, $d_T(\phi(i), \phi(i+1)) \leq 3, \quad \forall i$, where $d_T(i,j)$ is the distance between nodes i and j in the tree. They have also shown that $\sum_{i=0}^{|L|-2} d_T(\phi(i), \phi(i+1)) \leq 2(|L|-1)$, i.e., the average distance between adjacent nodes is less than 2. $|L|$ denotes the length of the loop. For any embedding of 2-dimensional arrays of n by n nodes in the leaves of a complete binary tree DeMillo, Eisenstat, and Lipton [23] have shown that there exist nodes (i,j) and $(i+1,j)$, adjacent in the array, such that $d_T(\phi(i,j), \phi(i+1,j)) > log_2 n - 3/2$. Rosenberg and Snyder [109] show that the average distance for the embedding of a 2-dimensional array in the leaves of a binary tree is at most $7 - 2^{-\lfloor log_2 n \rfloor + 1}$. Rosenberg and Snyder also consider the embedding of d-dimensional arrays with n^d nodes in the leaves of 2^d-ary and binary trees. The average distance between nodes adjacent in a d-dimensional array when embedded in a 2^d-ary tree is at most $4 - 2^{-\lfloor log_2 n \rfloor}$. The bound for a binary tree is $(4 - 2^{-\lfloor log_2 n \rfloor})d$. The maximum distance is at most $2d log_2 n$ for the binary tree embedding.

2.1.2 Boolean cube hosts

Nodes in a Boolean cube can be given addresses such that the addresses of adjacent nodes differ in precisely 1 bit. Furthermore, the number of adjacent nodes for any node equals the number of dimensions of the cube, i.e., the number of bits in the address. A **loop embedding** that preserves proximity is easily obtained for $|L| = 2^n$ by encoding the indices of the nodes in the loop in a *binary-reflected* Gray code [107]. Such Gray codes have several interesting properties. For instance, it is easy to show that $d_C(G_i, G_{(i+2^j) mod 2^n}) = 2$ for

Elem. index	Proc. index	Elem. index	Proc. index	Elem. index	Proc. index	Elem. index	Proc. index
0	0000	0	0000	0	0000	0	0000
1	0001	1	0001	1	0001	1	0001
2	0011	2	0011	2	0011	3	0011
3	0010	3	0010	3	0010	2	0010
4	0110	4	0110	7	0110	6	0110
5	0111	5	0111	6	0111	7	0111
6	0101	6	0101	5	0101	5	0101
7	0100	7	0100	4	0100	4	0100
8	1100	15	1100	12	1100	12	1100
9	1101	14	1101	13	1101	13	1101
10	1111	13	1111	14	1111	15	1111
11	1110	12	1110	15	1110	14	1110
12	1010	11	1010	11	1010	10	1010
13	1011	10	1011	10	1011	11	1011
14	1001	9	1001	9	1001	9	1001
15	1000	8	1000	8	1000	8	1000

Table 6: Conversion of Gray code to binary code

$j > 0$ [58]. This property is important for algorithms such as the FFT, bitonic sort, and cyclic reduction. For the FFT and bitonic sort an embedding according to a direct binary encoding of the indices of the data elements is preferable. However, application programs typically include the use of several different "elementary" algorithms, and a Gray code embedding may be preferable for other computations.

Another property of the binary-reflected Gray code is that for i even $d_C(G_i, G_{(i+3)mod2^n}) = 1$. Any loop of length $2^{n-1} + 2k, k = \{1, 2, ..., 2^{n-2}\}$ can be embedded in a n-dimensional cube (n-cube) such that $d_C(G_i, G_{(i+1)mod|L|}) = 1$ for $i = \{0, 1, ..., |L| - 1\}$ [50]. For $|L|$ odd there exists a node i in the loop such that $d_C(G_i, G_{(i+1)mod|L|}) = 2$. That the minimum maximum distance must be 2 is easily proved by considering the number of bit complementations in a cycle. In the following we refer to binary-reflected Gray codes simply as Gray codes.

An embedding according to the binary encoding of node indices in a loop does not preserve proximity. For i even $d_C(\phi(i), \phi(i+1)) = 1$, but for i odd $d_C(\phi(i), \phi(i+1))$ falls in the range $[2, n]$ ($d_C(\phi(2^{n-1} - 1), \phi(2^{n-1})) = n$). A Gray code encoding $G_i = (g_{n-1}, g_{n-2}, ..., g_0)$ can be rearranged to a binary encoding $i = (b_{n-1}, b_{n-2}, ..., b_0)$ in $n-1$ routing steps. The highest order bit in the Gray code encoding of an integer, and the highest order bit in its binary encoding coincide. The encodings of the last element, $N - 1$, differ in $n - 1$ bits. An element needs to be routed in dimension j if $g_j \oplus b_j = 1$. Routing the elements such that successively lower (or higher) order bits are correct yields paths that intersect at nodes only [50]. This form of routing amounts to reflections around certain "pivot" points in the Gray code. The pivot points are defined by the transitions in the bit being subject to routing. Table 6 illustrates the sequence of reflections that convert a 4-bit Gray code to binary code. A reflection consists of an exchange of elements between a pair of processors. Since each dimension is routed only once, no two elements traverse the same edge in the same direction during the entire process of data reallocation. If there are multiple data per node the routing of elements can be pipelined without conflict. This property is important, if a processor can concurrently support communication on all of its communication links.

The embedding of **d-dimensional meshes** with n_{d_i} nodes in dimension i is easily accomplished by partitioning the address space such that there are $\lceil log_2 n_{d_i} \rceil$ bits (dimensions of the cube) allocated for dimension d_i of the array. For $n_{d_i} = 2^k$ for some k this simple embedding is also efficient in the use of nodes in the cube. For meshes with sides that are not powers of 2 the embedding for most meshes can be made such that the expansion is minimum and the maximum dilation equal to 2 [43].

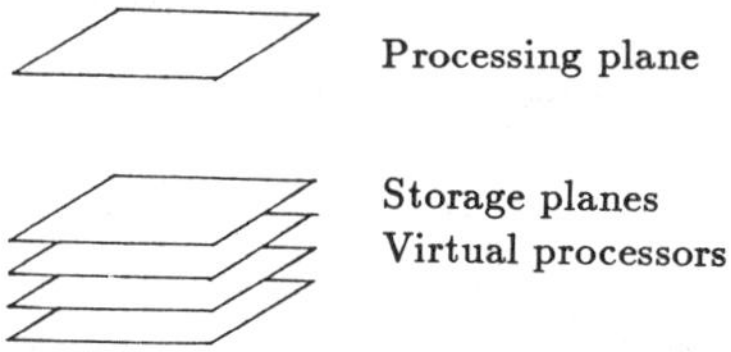

Figure 1: Processing and storage planes (virtual processors)

For the **FFT butterfly network** an identification of the corresponding nodes in different ranks with a cube processor yields a dilation 1 embedding. If $(x|y)$ is the address of a butterfly node, where x is n bits and y is $log_2 n$ bits, i.e., x gives the address within a rank of the butterfly, and y is the address of the rank, then all nodes with the same x are mapped into the same node of the Boolean cube with the scheme just suggested. Each cube node performs the task of $log_2 N$ butterfly nodes. Each butterfly communication is between nodes adjacent in the Boolean cube. If a binary-reflected Gray code encoding is applied to x, then butterfly communications are between nodes at distance two, except for the butterfly on the lowest order bit of x.

For the **data manipulator network** an identification of nodes of the network in the same way as for the FFT butterfly network does not yield communication between adjacent nodes, since any node communicates with nodes $i \pm 2^j$. One of these two communications are with an adjacent node, but the other is not, in general. However, if a binary-reflected Gray code encoding is applied to x, then proximity is preserved in that no communication is over a distance greater than two.

There exist many **spanning graphs** [44] in a Boolean cube. When buffer space is at a premium a Hamiltonian path may be the only choice, but if the data volume is low, then the height may be more important, and a *spanning binomial tree* [27,4] may be the best choice. For maximum utilization of the bandwidth n rotated and translated spanning binomial trees can be used. Such *edge-disjoint spanning binomial trees* [44] are optimum for large data volumes. In many instances several spanning trees are required concurrently, such as if all nodes broadcast data to all other nodes. If communication can take place only on one port at a time, then the spanning binomial tree routing is optimal, but if communication can take place on all ports concurrently, then routing according to *Balanced Spanning n-trees*, or *Rotated Spanning Binomial Trees*, is optimum [44].

2.1.3 Aggregation of data Elements

For an P by P matrix and a $2n$-cube with $P^2 > 2^{2n}$, elements of the matrix have to be identified, and stored in the same node of the ensemble. We consider two schemes of identifying matrix elements with nodes of a 2^n by 2^n array. In *consecutive* storage all elements $(i,j) = \{0,1,..,P-1\}$ of a matrix A are identified and stored in processor (p,q) $p = \lfloor \frac{i}{\lceil \frac{P}{2^n} \rceil} \rfloor$ and $q = \lfloor \frac{j}{\lceil \frac{P}{2^n} \rceil} \rfloor$. Each processor stores a submatrix of size $\frac{P^2}{2^{2n}}$. In *cyclic* storage the matrix elements are stored such that elements (i,j) are identified with node (p,q), $p = imod2^n$, and $q = jmod2^n$. In the *consecutive* storage scheme elements with the same *least significant bits* are identified with the same processor, whereas in the *cyclic* scheme the identification is made on the *most significant bits.*

With the consecutive storage scheme algorithms devised for the case of $P = 2^n$ can be employed with the apparent change of granularity. Operations on single elements are replaced by matrix operations. In the cyclic storage scheme the processing elements can be viewed as forming a processing plane, and the submatrices as forming storage planes, also known as virtual processors, Figure 1.

We find that the cyclic storage scheme enforces a greater insight into the communication and storage management issues. Elemental operations are of fine grain. For an ensemble architecture with communication overhead that is nonzero, or that is not proportional to the number of elements communicated, and that has pipelined arithmetic units, operations of fine grain should, in general, be merged for optimum use. Conversely, if the consecutive storage scheme is used it may be desirable to partition the elemental operations to increase the utilization of the ensemble. For matrix multiplication of square matrices there is no difference in processor utilization for the two storage schemes. But, for an algorithm such as LU-decomposition where the number of matrix elements involved in a step is decreasing throughout the computation, cyclic storage may be preferable. For LU-decomposition on a dense matrix it may yield a performance improvement in the range 1.5 - 2 [24]. For the solution of tridiagonal systems it yields a *performance degradation* [58]. The optimization of vector length, or communication packet size does not affect the optimum allocation of data, or the choice of algorithm.

2.2 Data Permutations

2.2.1 Conversion between storage schemes

Rearrangement of consecutive to cyclic storage order (or vice versa) can be carried out in time $P/2^n + n$ for P elements stored in a 2^n processor Boolean cube [50]. For this communication complexity it is required that a processor can support communication on multiple ports, and that the communication for successive stages can be pipelined. In the consecutive storage order the partitioning of the address space is $(a_{n-1}a_{n-2}\dots a_0|b_{m-1}\dots b_0)$, where the n highest order bits are processor addresses and the m lowest order bits are local addresses. In the cyclic storage scheme the bit fields are exchanged to $(b_{m-1}\dots b_0|a_{n-1}a_{n-2}\dots a_0)$. Clearly, if $m = n$ the storage conversion is equivalent to a matrix transposition. The exchange can be performed as a sequence of *shuffle* operations, i.e., left cyclic shifts on the address, or *unshuffle* operations through right cyclic shifts. Each such shuffle operation has a maximum path length of n edges. Performing the shuffle operations one at a time results in a communication time proportional to $min(n,m)n$. By performing a bit-wise exclusive-or operation on the addresses before and after the conversion it is clear that the maximum distance an element needs to traverse is n. Since each routing operation is an exchange operation and a dimension is only routed once, it follows that the paths are edge disjoint.

The rearrangement can be made recursively by a sequence of exchanges (exclusive-or operations) on distinct bits. The order in which the bits are treated is immaterial. All exchanges imply communication if $m \leq n$. An alternative implementation of the conversion algorithm is to perform exchanges of elements between pairs of processors differing in successively lower order address bits [50,41], and to perform local shuffle operations to make the elements that are being exchanged form a contiguos block. The processors with addresses in the lower half of the processor address space exchange the elements of the upper half of their local address space with the contents in the lower half of the local address space of their corresponding processors in the upper half of the processor address space. The result is that the first half of the processors contain the first half of the elements. An unshuffle operation on local addresses (or shuffle operation on the data) brings the first half of the data into row major order in the processors with addresses in the lower half of the address space, and the second half into row major order in the second half of the set of processors. The procedure is repeated recursively for each half independently, and concurrently.

Clearly, the local shuffle operation need not be carried out explicitly. Note that exchanges are always performed on half of the local address space, regardless of recursion step, or the number of rows or columns. This property is not true in forming the transpose of a rectangular matrix.

Carrying out the recursion in reverse order transforms a consecutive storage order to a cyclic storage order.

2.2.2 Matrix transpose

The formation of a matrix transpose is a particular permutation of data elements. With the matrix elements stored consecutively the encoding is $(rp_{n_r-1}rp_{n_r-2}\dots rp_0|rv_{m_r-1}\dots rv_0||cp_{n_c-1}cp_{n_c-2}\dots cp_0|cv_{m_c-1}\dots cv_0)$,

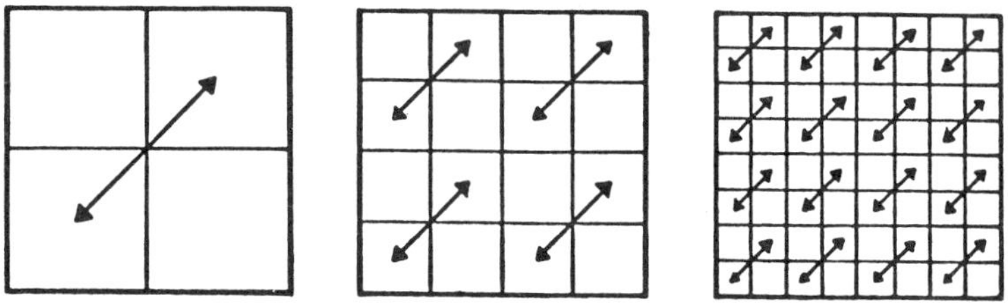

Figure 2: Recursive transposition of a matrix

where rp denotes the row processor addresses and rv the local addresses in the row direction (virtual row processors), cp the column processors and cv the local addresses for the column Direction. The transposition corresponds to exchanging the row and column bit fields. It can be carried out recursively [123,26,50,41] as illustrated in Figure 2 for $n_r + m_r = n_c + m_c$.

In the first step of the recursive procedure illustrated in Figure 2, the interchange of data is performed on the highest order bit of the row index *and* the highest order bit of the column index. In the second step the interchange is performed on the second highest order bit of the row *and* column indices, for all combinations of the highest order bits (i.e., 4 combinations). The number of index sets that differ in one bit of the row and column indices increases as the procedure progresses towards lower order bits.

For the consecutive storage scheme it is easily seen that with $n_r = n_c = n$, the first n steps imply interprocessor communication and with $m_r = m_c = m$ the last m steps are local to a processor node. These last steps consist of local address changes. (We presume here that the transpose is needed with some other data in some computation. Otherwise, the first n steps could also be accomplished without data movement by a suitable change of processor addresses.

With the cyclic storage scheme the situation is reversed. The first m steps amount to local address changes, whereas the last n steps require interprocessor communication. After the first m steps there are 2^{2m} matrices of size 2^n by 2^n to transpose. All matrices are stored identically.

With $n_r = n_c = n$ there is $2n$ dimensions to be routed. Indeed, all processors on the main anti-diagonal have elements that requires a routing distance equal to $2n$. With row and column dimensions taken pairwise all communications are exchanges over a distance of 2. The paths can be made edge-disjoint and communication pipelined. Moreover, constant storage per node suffices [50,41]. The element transfer time is $2^{2m}+2n-1$ accomplished by pipelining the 2^{2m} matrix transpositions, assuming that communication in both directions can take place concurrently on all of the ports of a processor. With the consecutive storage model and using the apparent granularity in the form of block operations the communication time is proportional to $2n \times 2^{2m}$, which for n large is considerably higher. The difference between the two expressions is due to the pipelining of element transfers in the first case. The same complexity is also attainable in the consecutive storage case by pipelining the transfers of elements of the blocks. It is possible to reduce the time further by establishing additional paths [41].

With a Gray code embedding of the array, successive row and column indices are always located in neighboring nodes of the cube. However, the communication required by the recursive procedure on row and column indices is between nodes storing elements of rows and columns whose binary encoding differs in successively higher or lower order bits. Each such communication requires the communication of elements in two dimensions, since complementing a bit in the binary encoding complements 2 bits in the Gray code encoding (except in complementing the least significant bit).

However, with $G(i)$ and $G(j)$ being the Gray code of the row index i and column index j the transpose operation for the case with $m_r = m_c = 0$ is equivalent to the communication implied by changing $(G(i)|G(j))$ to $(G(j)|G(i))$, which indeed is the same operation as in the binary encoded case. Routing row/column dimensions in descending order implies that matrix elements are subject to reflections around the main

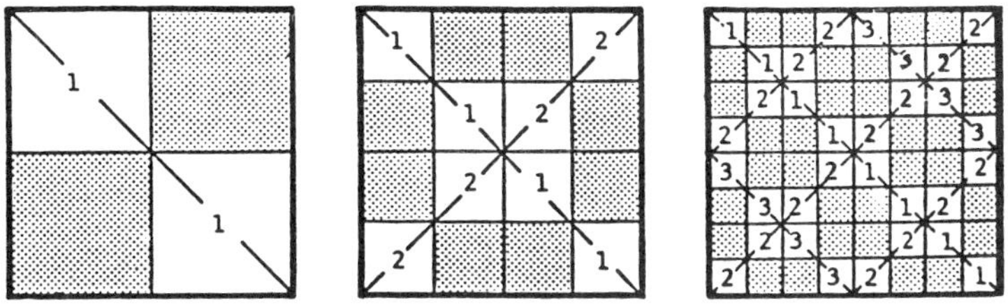

Figure 3: Transposing a matrix stored in a binary-reflected Gray code

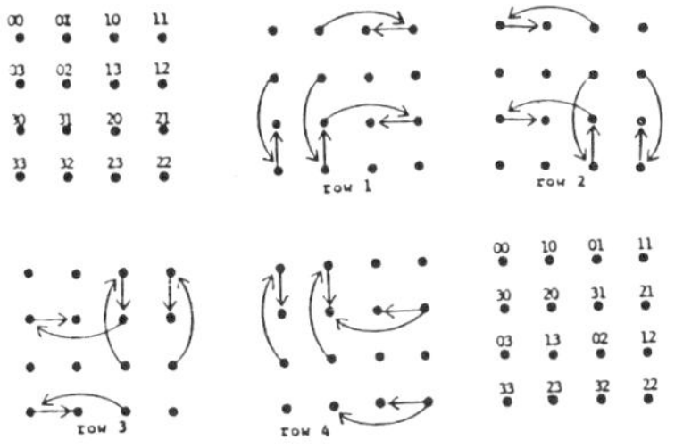

Figure 4: Routing paths in transposing a 4 by 4 matrix on a 4-cube

diagonal, and the anti-diagonal in alternating order. The behavior of the algorithm is illustrated in Figure 3. The numbers on the diagonals indicate the order of the reflection the submatrices are undergoing.

The application of the alternating descending order reflection algorithm to a 4-cube is illustrated in Figure 4. The routing algorithm can be made distributed.

Performing the transformation by a 2-dimensional mesh algorithm yields a considerably higher number of routing steps [50], $(2^{2m}+1)(2^{n-1}-1)/2$. The order of complexity of that algorithm cannot be reduced since $P(P-1)/2$ elements have to pass through $O(2^n)$ nodes, each of which has 4 ports.

The results presented for square matrices can be generalized to rectangular matrices [41].

2.2.3 Randomized Communication

Recently, several probabilistic algorithms of complexity $O(log_2 N)$ for arbitrary permutations on a Boolean n-cube and d-way shuffle networks have been devised. The probabilistic algorithms do not guarantee an even distribution of elements during permutation. The probabilistic algorithms have two phases. First, the elements are routed to a random location, then the elements are routed to the final destination. The routing is deterministic.

During routing from the initial location to the final destination during either phase, several elements may reach a node, then be delayed because of competition for a given communications link, even in the case that there is precisely one element per node in the initial and final states. Also, several elements may reside in a single node at the end of the first phase. Valiant and Brebner [131] show that for a Boolean cube with one element per node initially, and after the permutation, the queue length with high probability is at most of order $O(log_2 N)$. Indeed, for P/N elements per node they show that the probability that the permutation will require more than $(\alpha P/N+1)log_2 N$ routing steps is less than $(e/2\alpha)^{\alpha(P/N)log_2 N}$. They also

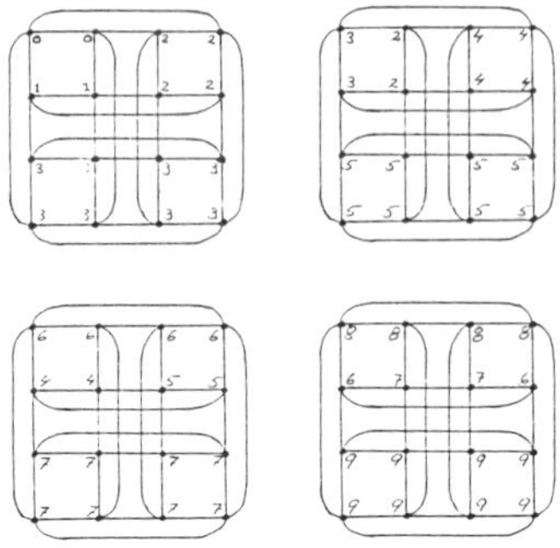

Figure 5: Spanning Binomial Trees in a Boolean cube, with sources in sequential order

establish similar bounds for so called d-way shuffle networks (in- and out-degree of a node is d), which also are considered by Upfal [130] and Aleliunas [3]. It is assumed that a processor can support communication on all its ports currently. Valiant and Brebner also show that for a n-dimensional mesh with $M = m^n$ nodes, the probability that at least one packet has not finished in time $(2n-1)(m + \alpha m^{3/4})$ is less than $C^{\alpha\sqrt{m}}$ for $C < 1$. This result compares favorably with the complexity of Batcher's bitonic sort or odd-even merge on meshes [128], [98], [79]. The Thompson and Kung algorithm yields a complexity of approximately $6\sqrt{M}$ for a 2-dimensional mesh, and $(3n^2 + n)M^{1/N}$ for a n-dimensional mesh. Simulations that exhibit a behavior well within the bounds for a variety of ensemble configurations are also presented.

2.2.4 Scan functions

Some operations applies to sets, such as broadcasting a value to a set of processors, finding the maximum or minimum in a given set of variables, or adding all the values. Critical issues in a system with distributed control (such as in a message passing system) are termination, completeness and uniqueness, i.e., that all nodes have received the message precisely once upon termination. Furthermore, local control of the distribution algorithm is desirable. In the Connection Machine, which is a bit-serial, synchronous, SIMD architecture, scan functions are available as operators in the programming language. The scan function requires that a spanning tree be generated for a specified set of processors. The encoding of the set of processors is most convenient if the processors form a contiguos domain in some index space, such as, for instance, a one- or multi-dimensional array. The encoding is particularly easy if the processors correspond to a specific bit-field. Such scans are known as segmented scans in the Connection Machine terminology [39]. The segments need not correspond to all possible addresses generated by a given bit-field, but should be contiguos for ease of encoding.

Assume for simplicity that the set of processors for which the scan operation shall be performed form a subcube of dimension k. Then, a spanning binomial tree for source node s is generated by every node i performing an exclusive or operation on its address and the source node address and communicating with all its neighbors corresponding to leading zeroes of $i \oplus s$. An interesting property of this simple scan routing is the following [50,53]. If the integers are embedded in the cube according to a binary-reflected Gray code, and the dimensions are routed in the order in which they first appear in the Gray code in going from i to $i-1$ in increasing order modulo the size of the subcube, then the order of arrival for every processor is the same as the order of scan initiation, if a scan operation is started at successive processors every other cycle,. This situation occurs in Gaussian elimination with pivoting on the diagonal.

The order preserving property for a binary-reflected Gray code is established by observing that the path i, $i+1$, $i+2$, ... reaches into 2 dimensions after 2 steps, 3 dimensions after 3 or 4 steps depending on i, and 4 dimensions after a minimum of 5 steps. The behavior of the algorithm for a 4-dimensional cube is shown in Figure 5.

2.3 Sorting

2.3.1 Combining sequential and bitonic sort on a Boolean cube

Stone [123] observed that the bitonic sort [5] maps well on to shuffle-exchange networks. From Stone's observations the implementation on a Boolean cube is immediate for one element per node. We will describe two algorithms for sorting P evenly distributed elements on a $N = 2^n$ processor Boolean cube for $P > N$.

The bitonic sort merges sorted sequences recursively. With one element per node the algorithm proceeds by comparison-exchange operations on elements that are located in nodes differing in 1 address bit, say the lowest order bit. Then, two sorted sequences stored in two 1-cubes are merged into one sorted sequence in a 2-cube. The sorting order, nonascending or nondescending, is determined by a mask. The mask is a function of the processor address and the length of the subsequences being merged. In all, $log_2 P$ sequences are merged serially. The number of sequences merged decreases from $P/2$ to 1. The final step merges two sequences stored in separate $n-1$-cubes into one sequence in an n-cube. The number of routing steps is $n(n+1)/2$, independent of the data. Each routing is performed in only one dimension. An algorithm for the bitonic sort expressed in pseudo code for $P = 2^n$ is as follows:

For $i := 1, 2, ..., n$ do
  If $i < n$ do
    nodes $a_{n-1}, .., a_{i+1}, a_i, a_{i-1}, .., a_0, a_i = 1$, set mask=1.
    nodes $a_{n-1}, .., a_{i+1}, a_i, a_{i-1}, .., a_0, a_i = 0$, set mask=0.
  end
  For $j := i-1, i-2, ..., 0$ do
    nodes $a_{n-1}, ..., a_{j+1}, 1, a_{j-1}, ..., a_0$, send their elements to
    nodes $a_{n-1}, ..., a_{j+1}, 0, a_{j-1}, ..., a_0$, which compare local
    and received elements
    nodes with mask=0 keep the smaller element and
    nodes with mask=1 keeps the larger
    rejected elements are sent to $a_{n-1}, ..., a_{j+1}, 1, a_{j-1}, ..., a_0$
  end
end

For $P = 2^n$ the last merge operation involves two sequences of length $P/2$. The merge is accomplished through a sequence of comparison-exchange operations on subsequences that decrease in length by a factor of 2 for each step. If $P > 2^n$, then additional sequential steps are necessary.

With the sorted sequence to be stored in cyclic order the first $log_2 P - n$ comparison-exchange operations of the final merge are local to a node, since the merge is performed recursively on successively shorter sequences, i.e., from the high order bits to the low order bits, and the higher order bits are local in the cyclic storage scheme. After these steps the result is $\frac{P}{N}$ bitonic sequences ordered with respect to each other. Each sequence has one element per node. The last n steps are separately performed on each of those sequences. Carrying out the first $log_2 P - n$ local steps as a bitonic merge yields poor performance. The operational complexity is $O(\frac{P}{N}(log_2 P - n)$, compared to $O(\frac{P}{N} + log_2 \frac{P}{N})$ for a sequential merge including bisection to find the maximum/minimum of the local bitonic sequence. This merge can be carried out concurrently in all processors [49]. The correctness of the algorithm can be proved by observing that the first $log_2 P - n$ steps, given the assumed storage order, realize N independent bitonic mergers, each for $\frac{P}{N}$ elements, and that corresponding output elements from these mergers form a bitonic sequence. The last n steps realize $\frac{P}{N}$ bitonic mergers for sequences of length N. The situation is a generalization of Batcher's construction [5] of a 16-sorter out of 4-sorters, see Figure 2.3.1.

The sorting is accomplished by recursively building longer sorted sequences, starting from the lowest order bits. The $\frac{P}{N}$ local elements belong to different sequences for the first n merges, each being a recursive merge. The time for cyclic sort by the algorithm outlined above is $T = \frac{P}{2N}(n(2log_2 P - n + 1)(4t_c + t_{ce})/2 + 2(log_2 P - n - 1)t_{ce}) + t_{ce}$, where t_c is the time for communication of an element between a pair of processors, and t_{ce} is the time for a comparison operation. If $P \gg N$ then T is of order $O((\frac{P}{N})log_2 P)$, and if $P \approx N$,

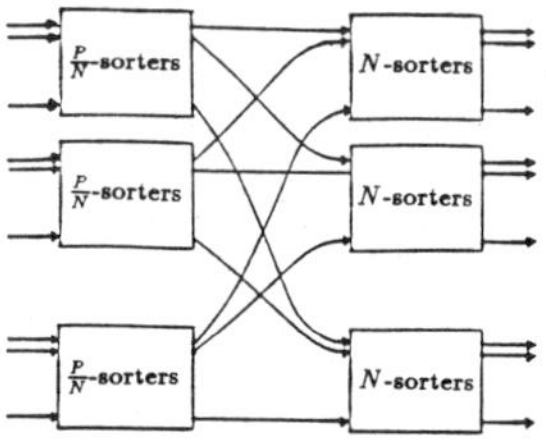

Figure 6: A network of N $\frac{P}{N}$-sorters followed by $\frac{P}{N}$ N-sorters.

T is of order $O(log_2^2 N)$. The speed-up is $O(N)$ for $N \ll P$ and gradually changes to $O(N/log_2 N)$.

With sorting into consecutive storage order and the elements initially stored consecutively, the first $log_2 P - n$ merges of the bitonic sort is local to a processor, with the merge sequence progressing from low to high order bits. These first steps generate a local sorted sequence, that is more efficiently created by a good sequential sort. The last n merges requires interprocessor communication, and involves sequences of $\frac{P}{N}$ elements instead of single elements. This algorithm is similar to the one proposed in [6]. The final $log_2 P - n$ steps of each of the last n bitonic merges are local to a processor and should be performed as a sequential merge. The communication and comparison complexity is of the same order as for sorting in cyclic order [49].

A cyclic storage order is generated by building sorted sequences over an increasing number of nodes first, then when a sorted sequence extends over all nodes include additional elements locally in the proper way. For the consecutive storage order the sorted sequences are first built locally, then extended over the processors when all local elements are included.

The running time of bitonic sort does not depend on the data distribution. This property is a drawback for nearly sorted sequences. The data movement in such instances can be reduced at the expense of additional logic for determining what subsequences should be exchanged. Such a modification can be made while preserving one advantage of bitonic sort, namely that the number of elements per node is kept constant during the sorting process.

2.3.2 Distribution counting

Rank assignment in the context of distribution counting [74] with L counters, or "buckets", can be carried out in a time of at most $[\frac{P}{N} + L + n - 1]2t_a + [L(1 - \frac{1}{N})3 + n]2t_c$ for $N \leq L$, and $[\frac{P}{N} + L + n - 1]2t_a + [6(L-1) + 5n - 3log_2 L]t_c$ for $N > L$, on a Boolean cube [49] (t_a is the time for an arithmetic operation). For few processors and a large number of elements compared to the number of buckets the algorithm offers linear speed-up. If the number of buckets is comparable to the number of elements to be sorted the speed-up is sublinear. For few buckets and few elements per node the speed-up is of order $O(N/log_2 N)$. The rank assignment algorithm that yields the complexity estimates above is data independent, as are the algorithms based on bitonic sort. The rank assignment algorithm is easily modified to deal only with non-empty buckets, which is efficient if only a few buckets in each node are populated. For particular distributions of elements the data dependent version will have a complexity of order $O(log_2 L)$ in the number of buckets, as in Hirschberg's shared storage model [40].

In the rank assignment algorithm multiple binary tree like computations are carried out concurrently. There are L binary trees with N leaf nodes each. The trees form subtrees of a tree with a total of $log_2 NL$ levels. Each node has a local copy of each bucket. To find the total number of elements in any bucket the number of elements in each of all the different copies of a bucket have to be added. This addition is carried out by the subtrees of $log_2 N$ levels. For the rank assignment partial sums are distributed from the root of the trees to the leaves. However, first an accumulation over all global bucket sums has to be performed. For each tree the summation/rank assignment process is carried out by recursive doubling [75].

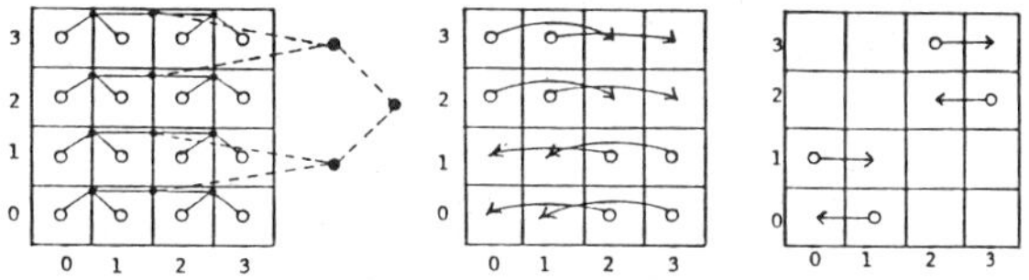

Figure 7: Concurrent tree computations on a Boolean cube

The recursive doubling process is carried out concurrently in all subtrees of height log_2N, by rooting the subtrees in different nodes of the ensemble. The global sums of different buckets are contained in distinct nodes. The tree embedding is such that one node in the Boolean cube contains $log_2N - 1$ non-leaf nodes of a subtree, another cube node $log_2N - 2$ non-leaf nodes of the same subtree, yet 2 other nodes $log_2N - 3$ non-leaf nodes of the same subtree, etc. After the first step half of the subtrees are treated by distinct halves of the cube. The divide-and-conquer process is repeated recursively. The speed-up for the subtrees of height log_2N is of order $O(N)$ for L of at least order $O(log_2N)$. The top log_2L levels of the tree are embedded similarly. Figure 7 illustrates the computations for $N = L = 4$.

2.4 Linear Algebra Computations

In this section we briefly describe a mesh algorithm by Cannon [16], which also is suitable for Boolean n-cubes since a two-dimensional mesh can be embedded in a Boolean cube with edge dilation one. We also briefly mention a few variations [50] of this algorithm. A recursive algorithm [21] that maps directly to a Boolean cube is also discussed, in particular its routing paths and pipelining properties. The multiplication of matrices of arbitrary shapes is treated in [66]. We also discuss some of the concerns in solving, dense, triangular, banded and tridiagonal systems on ensemble architectures and end with a discussion on the relative merits of FFT and tridiagonal solvers on such architectures.

2.4.1 Matrix multiplication

Cannon [16] presents an algorithm for computing the product C of two $\sqrt{N}$ by $\sqrt{N}$ matrices A and B stored in a 2-dimensional array of identical size. The algorithm requires $\frac{3}{2}\sqrt{N}$ communication steps, out of which $\lceil\frac{\sqrt{N}}{2}\rceil$ steps are for a set-up phase, and $\sqrt{N} - 1$ are for the multiplication phase. The purpose of the set-up phase is to align elements from the two matrices such that all nodes in the array can perform an inner product computation in every step in the multiplication phase. The alignment is accomplished by i cyclic shifts of row i of A and j cyclic shifts of column j of B. This skewing operation is the same as the alignment seen in many systolic algorithms [80,68].

The inner products defining the elements of C are accumulated *in-place*. Denote the storage cells for A, B and C by E, F and G. In the set-up phase the shifting yields: $E(i,j) \leftarrow E(i,(i+j)mod\sqrt{N})$, $F(i,j) \leftarrow F((i+j)mod\sqrt{N}, j)$, $G \leftarrow 0$ for $(i,j) \in \{0,1,2,...,\sqrt{N}-1\} \times \{0,1,2,...,\sqrt{N}-1\}$. Clearly $E(i,j) \times F(i,j)$ is a valid product for all i and j. In the multiplication phase the following operations are carried out: $G(i,j) \leftarrow G(i,j) + E(i,j) \times F(i,j)$, $E(i,j) \leftarrow E(i,(j+1)mod\sqrt{N})$, $F(i,j) \leftarrow F((i+1)mod\sqrt{N}, j)$, $i,j = \{0,1,2,...,\sqrt{N}-1\}$. With A a $P \times Q$ matrix and B a $Q \times R$ matrix the multiplication can be accomplished in a time of $max(\lceil\frac{P}{\sqrt{N}}\rceil, \lceil\frac{R}{\sqrt{N}}\rceil)\lceil\frac{Q}{\sqrt{N}}\rceil(\sqrt{N}-1)t_c + \lceil\frac{P}{\sqrt{N}}\rceil\lceil\frac{Q}{\sqrt{N}}\rceil\lceil\frac{R}{\sqrt{N}}\rceil((\sqrt{N}-1)(t_a + max(t_a,t_c)) + 2t_a)$.

A drawback of the algorithm by Cannon is that no computations are being performed during the alignment process. Some elements make almost 2 full revolutions, should only unidirectional communication be allowed. However, one revolution suffices, and algorithms can be devised such that successive matrix multiplications can be initiated every $\sqrt{N}$ "cycles". For instance, using the outer product formulation [47] of a matrix product, and passing the columns of A along rows (one element per row) in order of increasing column indices, and rows of B along columns in the direction of increasing row indices. The distribution of columns of A and rows of B can start from the locations where the elements are stored [54]. The distribution can be pipelined, and the initiation of the distribution of the different columns and rows can be spread over time in order that no temporary storage be needed, other than for a pair of elements to be multiplied. With only unidirectional communication, and end-around connections, a total time of $5(\sqrt{N}-1)$ "cycles" is required for one matrix multiplication. The complexity of the algorithm may be improved [50], but with unidirectional data movement pipelining is easy to visualize. The data movement is similar to that of the dense matrix factorization algorithm (without partial pivoting) described later.

The complexity of the algorithm can be reduced by a term $\sqrt{N}-\frac{1}{2}n$, if the alignment can be accomplished in time n instead of $\sqrt{N}$. For matrices of a size comparable to the number of processors this difference is also significant relative to the total time. Dekel et. al. [21] describes such an algorithm for $\sqrt{N}$ by $\sqrt{N}$ matrices embedded in a Boolean cube of N nodes by a separate binary encoding of row and column indices. The algorithm has a set-up phase in which A and B are arranged such that $E(i,j) \leftarrow A(i, i \oplus j)$ and $F(i,j) \leftarrow B(i \oplus j, j)$. Hence, $E(i,j) \times F(i,j)$ are valid terms for $C(i,j)$ for $(i,j) \in \{0, ..., \sqrt{N}-1\} \times \{0, ..., \sqrt{N}-1\}$. The rearrangement requires exchanges of elements in the dimensions specified by i for A and by j for B. Clearly, the set-up phase requires n steps, and no two elements traverse the same edge in the same direction. The set-up phase for multiple multiplication operations can be pipelined so that the total set-up time for P problems is $P+n-1$.

In the multiplication phase nodes exchange their content in an order determined by the transition sequence of the bits in a binary-reflected Gray code [21]. It follows that the time for multiplying a Q by R matrix B by an P by Q matrix A on a Boolean n-cube, with A, B and $C \leftarrow A \times B + C$, embedded according to a separate binary encoding of row and column indices, is at most $((\lceil \frac{P}{\sqrt{N}} \rceil + \lceil \frac{R}{\sqrt{N}} \rceil) \lceil \frac{Q}{\sqrt{N}} \rceil + log_2 N - 1) t_c + \sqrt{N} \lceil \frac{P}{\sqrt{N}} \rceil \lceil \frac{R}{\sqrt{N}} \rceil \lceil \frac{Q}{\sqrt{N}} \rceil max(2t_a, t_c)$. The number of submatrices of size $\sqrt{N} \times \sqrt{N}$ is $(\lceil \frac{P}{\sqrt{N}} \rceil + \lceil \frac{R}{\sqrt{N}} \rceil) \lceil \frac{Q}{\sqrt{N}} \rceil$. The number of block matrix multiplications is $\lceil \frac{P}{\sqrt{N}} \rceil \lceil \frac{R}{\sqrt{N}} \rceil \lceil \frac{Q}{\sqrt{N}} \rceil$. Cannon's matrix multiplication algorithm is devised for SIMD architectures. For mesh or Boolean cube configured ensembles of the MIMD type it is possible to devise algorithms with many different kinds of data flow and a complexity of $(\lceil \frac{P}{\sqrt{N}} \rceil \lceil \frac{R}{\sqrt{N}} \rceil \lceil \frac{Q}{\sqrt{N}} \rceil \sqrt{N} - 1) max(2t_a, t_c) + \alpha d + 2t_a$, where d denotes the diameter of the ensemble configuration and $\alpha \leq 4$ [54].

The multiplication of rectangular matrices is treated in detail in [66]. Depending on the shape of the matrices and the parameters of the machine, any matrix algorithm as outlined above, or a matrix-vector algorithm as described below, or the computation of the transpose of the product may be the optimum.

2.4.2 Multiplication of a full matrix by a triangular matrix

If A is an upper triangular (or lower triangular) N by N matrix, then only half of the arithmetic operations $N(N+1)/2$ are nontrivial. The alignment and multiplication phases of Cannon's algorithm can be interleaved such that computations start from one corner of the array and progress towards the opposite corner. The total data movement is the same. In this variation of Cannon's algorithm it is convenient to use the notion of *computational windows.* A computational window is defined by the data elements processed concurrently by the ensemble nodes. The computational window during step j for an upper triangular matrix A, $a_{ij} = 0$ for $i - j > 0$, is shown in Figure 8. It also shows the computational window for step j, if A is a strict lower triangular matrix, $a_{ij} = 0, ij \leq 0$.

From Figure 8 it is obvious that the multiplication $A \times B$ and $D \times B$, where A is upper triangular and D strict lower triangular of dimension $\sqrt{N}$ by $\sqrt{N}$, or A strict upper triangular and D lower triangular can be performed concurrently on a torus of dimension $\sqrt{N}$ by $\sqrt{N}$, or a Boolean n-cube.

2.4.3 Matrix-vector Multiplication

The matrix multiplication algorithms described above can also be used for matrix-vector multiplication, $Y = AX$. However, the running time is independent of the number of columns of X, and the data movement is larger than necessary [54]. An algorithm that on a Boolean cube is of a lower complexity than the

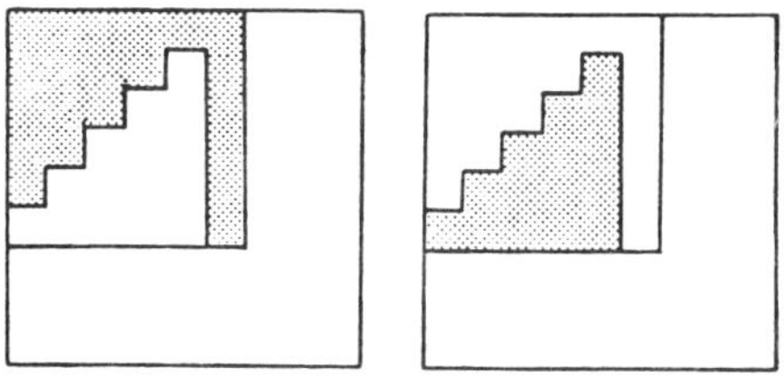

Figure 8: The computational window at step j for computing $C \leftarrow A \times B + C$, A upper triangular, or strict lower triangular

algorithm by Cannon (adapted to a Boolean cube), or the algorithm by Dekel et. al., for a single vector, or for a matrix X with few columns is obtained by making A stationary, distributing the elements of X to the proper ensemble nodes, and accumulating the partial products over space to yield C in the desired location.

To outline the algorithm assume A is a $\sqrt{N} \times \sqrt{N}$ matrix and x a $\sqrt{N}$ vector with components x_i. Assume that the vector x is aligned with the first column of A. First x_i is rotated i steps in the direction of increasing column index for $i = \{0, 1, ..., \sqrt{N} - 1\}$, then each x-value is distributed to all nodes in column i, and the products computed. Finally, the products are accumulated. Each of these steps can be carried out in a time proportional to n. With the matrix embedded by separately encoding row and column indices in a Gray code, the shifting is performed in different subcubes, and no communication conflicts occur. The routing of elements for a given shift s can be carried out by comparing the Gray codes of i and $i + s$ and moving towards the desired address by one dimension at a time in any order. A copy-scan can be used within columns. A sum-scan can be used for the accumulation of inner products. Complexity estimates for algorithms computing matrix-vector products by accumulating inner products *in-space* are given in [54] for dense matrices, and in [55] for banded matrices.

2.4.4 Factorization of dense matrices

The algorithms for Gaussian elimination and Gauss-Jordan elimination described below can be viewed as modifications of systolic algorithms [80], [52]. The modification of the symmetric versions of Gaussian elimination such as Cholesky's, Crout's, and Doolittle's methods can be carried out in a similar way. Systolic algorithms for mesh configured ensembles for Cholesky's method are given in [2], [63], for Given's rotations in [30], [33], [2], [59], and for Householder transformations in [51]. Given's and Householder's methods make use of unitary transformations and are numerically stable.

The factorization of a matrix A into a lower triangular matrix L and an upper triangular matrix U, is carried out such that the product form of L^{-1} is computed, $L^{-1} = L_{N-1}L_{N-2}...L_1$. $A = LU$ and $Ux = L^{-1}y$. The elements of the factors are stored in the same locations as the elements of the matrix to be factored.

The non-trivial elements of a factor become known after the preceeding factors have been applied to A, i.e., $l_{ki}, k = \{i, ..., N-1\}$ equals the corresponding elements of $A_i = L_{i-1}A_{i-1}, A_0 = A$. The application of the factors can be pipelined, as is done in systolic algorithms. In Gauss-Jordan elimination the inverse is also expressed in product form, $A^{-1} = J_{N-1}J_{N-2}....J_0$. The non trivial column of J_i is determined by the corresponding column of $A_i = J_{i-1}A_{i-1}$.

We assume that A is stored in a 2-dimensional array with one element per processor. We first present algorithms for 2-dimensional arrays, then describe modifications that can take advantage of the added communication capability of a Boolean cube. In the application of L_i to A_i row i (the pivot row) is distributed to rows k, $k > i$, and column i to columns l, $l > i$. In Gauss-Jordan elimination the pivot row is distributed to all other rows. If the matrix is of the same dimension as the array, i.e., there is only one matrix element per node, then Gauss-Jordan elimination can be completed in the same time as Gaussian

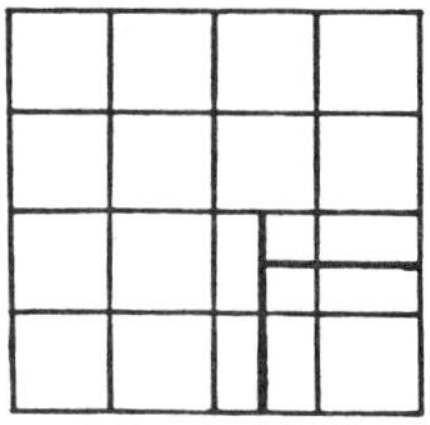

Figure 9: Storage and distribution of pivot row and column in dense matrix factorization

elimination. An increasing number of processors become idle in Gaussian elimination.

For A large compared to the array only the diagonal blocks are diagonalized, with A being stored cyclicly. The storage of the pivot row and columns [54] and their distribution is illustrated in Figure 9. Each application of a factor is similar to performing a column by row product in the outer product matrix multiplication algorithm.

For each column elimination operation a number of elements need to be distributed along rows, and a number along columns. The number of elements distributed along rows equals the number of submatrices on and below the diagonal. The number of elements distributed along columns is equal to the number of blocks on, and to the right of the diagonal, including the right hand sides.

A Boolean cube offers a capability of carrying out the distribution of the pivot row to other rows and the pivot column to other columns in less than linear time. For the factorization of a matrix by Gaussian elimination without partial pivoting this capability does not lower the complexity of the elimination, but it does in the event of partial pivoting. However, in forward substitution on multiple right hand sides, and in Gauss-Jordan elimination the communication capability of the Boolean cube can be used to reduce the complexity of the propagation term. The order of the complexity is still linear in the size of the matrix, which is intrinsic to Gaussian elimination without partial pivoting. Faster methods for band matrix problems are described in the next section.

In performing the data distribution in Gaussian elimination, or Gauss-Jordan elimination, without partial pivoting the source nodes are consecutively indexed. A correct result is guaranteed, if data arrives in the same order as its distribution is initiated. This condition is sufficient, but not necessary for correctness. The scan algorithm described earlier guarantees the same order of arrival as that of distribution [55]. Note that in Gaussian elimination the distribution for the last several equations only needs to cover successively smaller cubes.

2.4.5 Solution of Triangular Systems

A number of methods for the solution of general linear recurrences $Lz = y$ (y and z are vectors, L a lower triangular matrix) on architectures with global storage have been proposed, and their complexity analyzed, assuming zero communication cost. Sameh [115] gives a survey of such algorithms and their properties. We present an algorithm for mesh or Boolean cube configured ensembles [50]. It is an adaptation of the binary tree algorithm by Johnsson [57], which in turn is a particular instantiation of the column-sweep algorithm described by Kuck [77]. We assume that the vectors y and z are stored in row major order, and L in column major order (L^T is stored in row major order). In the algorithm outlined below y and z are stationary, L communicated along rows, and partial inner products along columns.

The elements of a column of L are passed along rows of the array. The elements of a column are passed in order of increasing row index. The first element of a column of L is used to compute a new component of z. Subsequent elements of a row of L^T are multiplied by this z component, added to the corresponding partial inner product passed along columns in direction of increasing row indices, and the result passed to

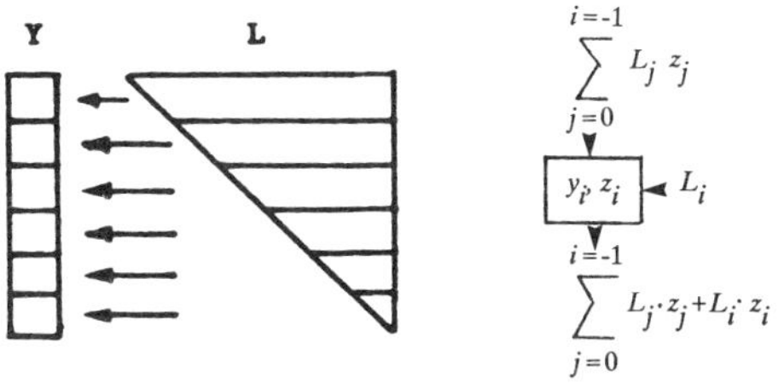

Figure 10: Data movement in a linear recurrence algorithm on a torus

the next processor in the same column. The first partial inner product that reaches a processor is used to update the right hand side, before a new z is computed. Hence, a processor in row i when first activated computes z_i, then computes the product $l_{ki}z_i$, adds this product to $\sum_{j=0}^{i-1} l_{kj}z_j$ received from the preceeding row, and outputs the result to the succeeding row.

This algorithm for solving linear recurrences progresses from one row to the next at the rate $t_d + 2t_a$, ignoring communication time (t_d is the time for division of two floating-point numbers). For a 2-dimensional array of $\sqrt{N}$ by $\sqrt{N}$ processors, the service of a processor is requested for a new row every $(t_d+2t_a)\sqrt{N}$ units of time. If L is a banded matrix with m nonzero diagonals, then a processor needs a time of $t_d + 2(m-1)t_a$ to complete the computations for one column of L. The time to solve the linear recurrence by this algorithm is approximately

$$\lceil \frac{P-m}{\sqrt{N}} \rceil max(t_d + 2(m-1)t_a, (t_c + t_d + 2t_a)\sqrt{N}) + p \sum_{j=0}^{\lfloor \frac{m}{\sqrt{N}} \rfloor} max(t_d + 2(m - j\sqrt{N} - 1)t_a, (t_c + t_d + 2t_a)\sqrt{N}).$$

For banded systems a recurrence solver can also be based on the partitioning method [116]. This approach can further reduce the complexity of solving linear recurrences. The partitioning method is discussed further in the next section.

2.4.6 Banded System Solvers

Tridiagonal systems

Irreducible tridiagonal systems of equations of order P can be solved in $2log_2P$ steps using $O(P)$ arithmetic operations by odd-even cyclic reduction [15]. The method has been modified by Hockney [47] to yield a solution in log_2P steps, but at the expense of $O(Plog_2P)$ arithmetic operations. For highly concurrent ensembles it is of interest to find mappings of the computation graph on to the nodes of the ensemble such that the communication complexity is no higher, or at least of the same order as the parallel arithmetic complexity. Binary trees, shuffle-exchange networks, and Boolean cubes allow for global communication in a time proportional to p for $P = 2^p - 1$ and $P = 2^p$ processors respectively.

The solution of tridiagonal systems on binary trees is interesting not only for the importance of efficient tridiagonal solvers, and the relative simplicity of constructing large tree ensembles, but also from an algorithm design point of view. There exists a mapping of the computation graph for cyclic reduction on P equations on to a binary tree of P nodes such that the communication complexity is $3log_2P$ [58]. A comparable communication complexity is also obtainable on shuffle-exchange networks and Boolean cubes [58].

The computation graph of cyclic reduction is shown in Figure 11. For $P = N$ mapping the equations on to nodes in the tree in inorder for every level of the computation graph, yields a map with the desired order

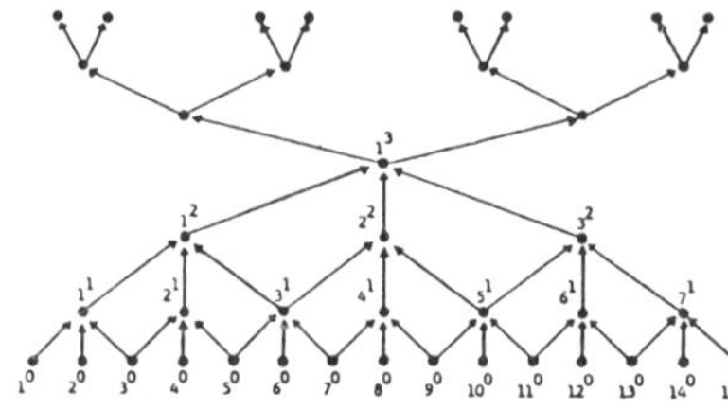

Figure 11: The computation graph for odd-even cyclic reduction

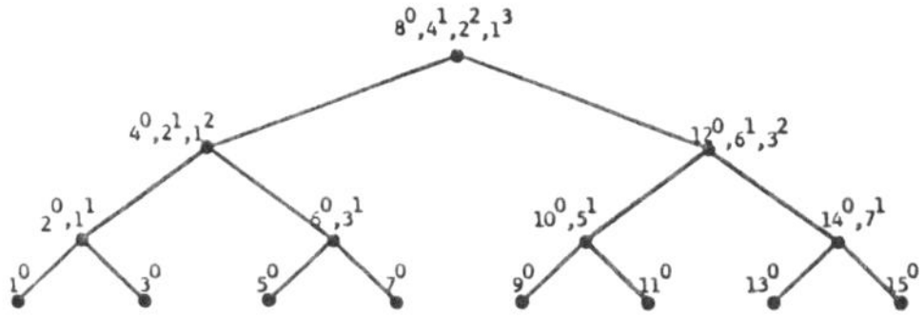

Figure 12: Inorder mapping of cyclic reduction on to a binary tree

of complexity. The first reduction step requires a time proportional to n, the second a time proportional to $n-1$, etc. But, the reduction steps can be pipelined, and the total time is proportional to $3n$ [58], [28]. The inorder mapping is shown in Figure 12.

For $P > N$ several equations must be identified with the same processor node. For load balancing it is desirable to make the division as even as possible. For simplicity we assume here that $P = 2^m N$, i.e., that m address bits are required for the local address. The consecutive storage scheme can be considered as forming a quotient graph from the computation graph by combining a successively indexed nodes at each level of the computation graph into a node in the quotient graph. This approach is similar to domain decomposition in the solution of partial differential equations. The nodes at each level of the quotient graph are then mapped on to the processor tree in inorder. The number of quotient nodes at the leaf level of the computation graph with $\lceil \frac{P}{N} \rceil$ equations is $2^{pmodn-1}$, which corresponds to a *pmodn* level binary tree. In the formation of the quotient nodes the computation graph is effectively partitioned into "vertical" slices, with one quotient node per slice and level, for $p - n$ levels starting with the leaf level. The quotient graph approach provides the best possible computational balance.

A critical observation in finding a communication efficient mapping is that the communication between some pair of partitions alternates in direction for every level (reduction step) of the computation graph. The efficiency of the inorder map relies on the fact that the communication is unidirectional, and can be pipelined. Hence, odd-even cyclic reduction applied to this mapping is not efficient, but by changing the elimination order to substructured elimination the mapping is effective. Only one communication is required in the substructuring phase. The amount of fill-in is approximately the same as in odd-even cyclic reduction, and so is the arithmetic complexity. The reduced system is then solved by cyclic reduction using an inorder map. The total complexity is of order $O(\frac{P}{N} + n)$ [58,62].

The substructured algorithm is of minimum order of complexity, both with respect to communication and arithmetic. For a diagonally dominant system the the diagonal dominance increases during substructuring [108] and no or only a few reduction steps may be sufficient. If a few cyclic reduction steps suffice, then a proximity preserving embedding [109] of the quotient graph may be advantages. The reduction in computational complexity accomplished by truncating the reduction process is relatively much more significant in a highly concurrent system than in a single processor system. For $P = N$ the running time is proportional to the reduction steps executed, while on a single processor half of the total (untruncated) execution time is

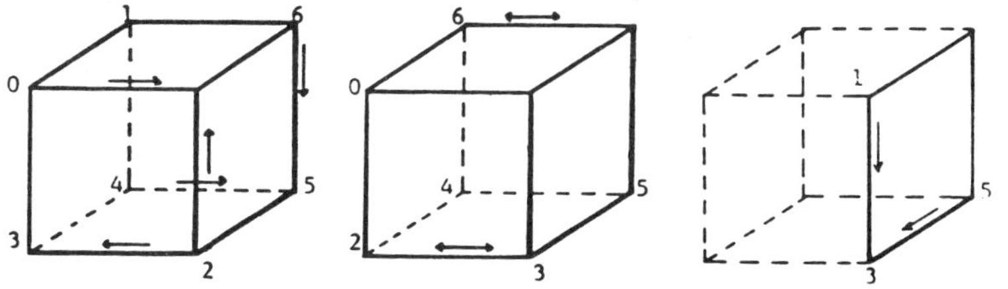

Figure 13: Cyclic reduction on a Boolean cube

spent in the first reduction step, a quarter in the second, etc. The speed-up for cyclic reduction and $P = N$ is $O(\frac{N}{log_2 N})$, but approaches $O(N)$ if the reduction process can be terminated after a fixed number of steps, as in strongly diagonally dominant systems.

On a Boolean cube substructured elimination with odd-even cyclic reduction for the reduced system has a communication complexity $O(n)$. Each step of the cyclic reduction algorithm involves 3 nodes of the computation graph. An embedding according to a binary encoding would require communication across n edges for the first step of the algorithm, $n - 1$ edges for the $2nd$ step, $n - 2$ for the $3rd$ step, etc. The binary-reflected Gray code also allows for simple, distributed control. Each processor can determine with which neighboring processor to communicate, and what information shall be transmitted/received from its address, and the reduction step currently being executed [58,62]. Because of the properties of the binary-reflected Gray code each step requires only 2 routing steps. One of these routing steps can be carried out as an exchange operation (but need not be). In such a case successive levels of the computation graph are mapped into subcubes of monotonely decreasing dimensionality. For $P = N$ all processors participate in the first reduction step, about half of which only perform communication. The equations participating in the second reduction step are moved to one half of the cube, and the process is repeated recursively. Figure 13 illustrates a few steps in the reduction process for $P = N = 8$.

Odd-even cyclic reduction has a higher arithmetic complexity than Gaussian elimination, and it also requires more communications than if a transposition of the data is performed to one processor, and the result distributed back to where the equations came from. Hence, for certain combinations of arithmetic and communication capabilities it may be faster to use a transposition and a sequential algorithm, and even to avoid substructuring [113,67].

If multiple independent tridiagonal systems are to be solved, then either all problems can be distributed over the entire ensemble, or the ensemble can be logically partitioned such that each problem is solved by a partition. For tridiagonal systems and the solution methods discussed here, it is always advantageous to partition the ensemble, even in the event of negligible communication time [58].

A detailed experimental study of optimum methods for the solution of single and multiple tridiagonal systems on the Intel iPSC is reported in [67].

General Banded Systems

The substructuring technique is has also been applied to banded systems. Sameh and collaborators [116,83,25] use partitioning to reduce banded systems of bandwidth $2m + 1$ to dense, block pentadiagonal, systems of order $2mN - 1$ for N partitions. The blocks are of size $m \times m$. The solution of the reduced system by Gaussian elimination on a linear array is considered in [83], and the solution by block-Jacobi and preconditioned conjugate gradient methods in [25]. Reiter and Rodrigue [108], and Johnsson [61,60,24] analyze a slightly different substructuring that reduces the banded system to a dense, block tridiagonal system of order mN. Reiter and Rodrigue give conditions under which diagonal dominance is preserved during the Gaussian elimination part of the algorithm. Johnsson shows that the arithmetic complexity in deriving the tridiagonal system is approximately 1/3 of that required in deriving the pentadiagonal system, and analyze the complexity of solving the reduced system by Gaussian elimination and block cyclic reduction

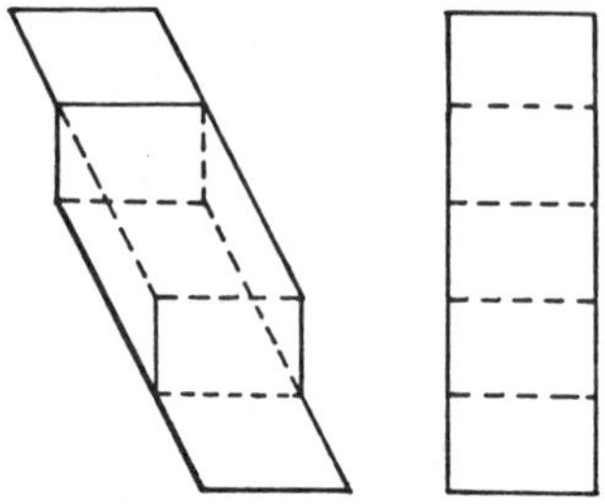

Figure 14: Band matrix factorization on a 2-dimensional array or Boolean cube

on linear arrays, binary trees, shuffle-exchange networks and Boolean cubes.

The optimum number of partitions N_{opt} depends on m, P, and the ratio of the communication and computation bandwidths. N_{opt} for Gaussian elimination on a linear array is of order $O(\sqrt{\frac{P}{m}})$ and the corresponding complexity is of order $O(m^2\sqrt{Pm})$. Block cyclic reduction yields a lower complexity under a variety of conditions, even on a linear array. The value of N_{opt} falls in the range $O(\sqrt{P}) \leq N_{opt} \leq O(\frac{P}{m})$, and the corresponding complexity is in the range $O(m^2\sqrt{P} + m^3 log_2 P)$ to $O(m^3 + m^3 log_2 \frac{P}{m})$ [61]. For binary trees, shuffle-exchange networks, and Boolean cubes N_{opt} is of order $O(\frac{P}{m})$ and the corresponding complexity of order $O(m^3 + m^3 log_2 \frac{P}{m})$. For small matrix bandwidths this algorithm yields good speed-up, but as the bandwidth increases the speed-up becomes low.

The above results apply under the assumption that there is one processor per partition. The number of partitions is constrained to be at most $\frac{P}{m}$. However, it is possible to exploit concurrency in the operations also within the partitions, which for m of order $O(P)$ is the main source of concurrency. For instance, one can use $N_c \leq m^2$ processors configured as a mesh or a Boolean cube during the elimination of the elements in one column, as in systolic algorithms [80] [52], but use the dual formulation in which the factors are computed *in-place*. The speed-up is of order N_c. The computations proceed in two phases: factorization with forward substitution, and backsubstitution. The factorization can proceed from the first to the last column, or from both ends concurrently. In the latter case an m by m dense system of equations must be solved for the "middle" equations before backsubstitution takes place. A dense m by m system is also solved in the 1-way elimination scheme (the last m equations). The backsubstitution consists of solving a linear, banded recurrence. The technique discussed previously can be used. Figure 14 illustrates an intermediate state of the factorization process.

The scan algorithm can be used to reduce the propagation time for a Boolean cube, and multiple right hand sides. The propagation time then becomes of lower order even in the case of $N_c = m^2$, and $P \approx m$. The complexity of solving banded systems by this approach is $O(m^2 \frac{P}{N_c})$ $(N = 1)$ [55]. For symmetric matrices storage as well as time can be saved using a parallel version of Cholesky's method [2], [63].

The two methods can be combined such that N_c processors are used for each partition. Such a set of processors is referred to as a *cluster*. With $\frac{N}{N_c}$ clusters of N_c processors each, intracluster connections in the form of a 2-dimensional mesh (or torus) or Boolean cube, and intercluster connections forming a binary tree, shuffle network, or Boolean cube, the minimum time complexity is of order $O(m + mlog_2 \frac{P}{m})$, and N_{opt} of order $O(N_c \frac{P}{m})$, and $N_{c_{opt}}$ of order $O(m^2)$ [55]. Note that for a Boolean cube a subcube can be considered as a cluster.

The combined algorithm degenerates to the simple band matrix algorithm proceeding along the band from one corner to the other for $m = P - 1$. For $m = 1$ it degenerates to the tridiagonal solver described previously.

2.4.7 Fast Poisson Solvers

Fast Poisson solvers combine Fast Fourier Transforms (FFT) with tridiagonal system solvers, and block cyclic reduction to achieve a minimum arithmetic complexity of order $O(Plog_2log_2P)$ for a $P \times P$ grid [45], [46], [124]. Stone [123] observed that the FFT can be carried out in log_2P steps on a P-node shuffle-exchange network. The modification of Stone's algorithm for a Boolean cube is straightforward. Shuffle operations become unnecessary. The butterfly operations are simply carried out on elements residing in processors adjacent in different dimensions. This property holds for decimation-in-time (DIT) as well as decimation-in-frequency (DIF) FFT.

With multiple elements per processor the initial (and transformed) sequence can be stored in either consecutive or cyclic storage order. In either case, and depending on whether a DIT or a DIF FFT is used the first, or the last, $log_2\frac{P}{N}$ butterfly operations are local to a node. The arithmetic complexity is of order $O(\frac{P}{N}log_2P)$ and the communication complexity is of order $O(\frac{P}{P}log_2P)$. The speed-up is proportional to P. FFT algorithms for linear arrays are given in a number of references [104], [68], [64]. An early description of a FFT on a 2-dimensional array is given by Stevens [122]. An analysis of the area-time aspects of the FFT on a variety of ensemble configurations is given in [127].

The solution of Poisson's problem on a rectangle can be obtained by a 2-dimensional FFT, by a number of 1-dimensional FFTs that decouple the equations into a set of independent tridiagonal systems, or by a combination of block cyclic reduction, FFT, and tridiagonal system solvers, and the so called FACR method [45], [46], [15], [14], [124], [125], [126]. By exploiting symmetries and using real transforms the number of arithmetic operations per point is less than $2.5log_2P$ in the FFT computation. Using Gaussian elimination with precomputed factors [125] the number of arithmetic operations per point for the tridiagonal solvers is 4, or approximately 4 if advantage is taken of the fast convergence of the elements of the factors [99]. Hence, the arithmetic operations count is less for solving the tridiagonal systems by Gaussian elimination than by FFT, solution of a diagonal system, and inverse FFT (IFFT). A cyclic reduction algorithm could be used, but the operations count per point with precomputed factors is 6.

In a highly concurrent system the differences in complexity between the FFT and tridiagonal system solvers are much smaller. To amplify this issue consider the case with P^2 nodes in an ensemble configured as a Boolean cube. The solution of Poisson's equation either by a 2-dimensional FFT, or by a combination of 1-dimensional FFT's and tridiagonal system solvers based on cyclic reduction, then requires a time of order $O(log_2P)$, including communication. For this extreme case Gaussian elimination is not of interest with respect to computational complexity, since it is inherently sequential. The number of communications in the Boolean cube is $2log_2P$ for odd-even cyclic reduction on P equations, which can be reduced to log_2P for parallel cyclic reduction [47]. Even though it seems preferable to use parallel cyclic reduction, this is not necessarily true [56]. With a binary-reflected Gray code embedding of the equations each such communication, but the first is over two edges. In a packet switched communication system the number of communications is $4log_2P$ and $2log_P$, respectively. It suffices with $4log_2P$ communications even if the communication system only supports one send *or* one receive operation at a time. In the case of the Poisson equation only the right hand side need to be transferred. The number of real arithmetic operations is at most $7log_2P$, which can be reduced to $6log_2P$ with precomputed factors.

With P^2 processors a real FFT on P points requires $5log_2P - const$ operations per point, if the butterfly computations are split between two processors, otherwise $10log_2P - const$. In the former case two exchanges are required per butterfly, otherwise one exchange suffice. If the communication system only supports one send or one receive operation, then each exchange requires two communications. With a binary encoding of the lattice all communications are over single edges, but if the lattice is embedded by a binary-reflected Gray code embedding then the communication is over two edges. Hence, an FFT and an inverse FFT on P points require between $2log_2P$ and $16log_2P$ communications depending on the communication system, the embedding, and the load balancing. Each communication involves a complex variable.

Using a tridiagonal solver instead of an FFT-IFFT saves at least $3log_P$ to $4log_2P$ arithmetic operations and maybe also communication (depending upon the architecture). In the FFT-IFFT approach all nodes are used in all steps, but in the cyclic reduction tridiagonal system solver the number of active nodes decreases. Most of the tridiagonal systems are sufficiently diagonally dominant that the reduction process can be truncated. This property does not reduce the total solution time in this extreme case (P^2 processors for P^2 lattice points). With fewer processors it gives rise to an interesting load balancing problem.

Sameh [114] presents a method for the solution of the 2-dimensional Poisson equation on a ring of processors, and for the solution of the 3-dimensional problem on a cylinder of processors. In the 2-dimensional case FFT's are performed on data local to a processor, and the tridiagonal systems solved by a modification of the partitioning method [116]. The modification is made to take advantage of the Toeplitz form of the tridiagonal matrices. The reduced systems are solved by pipelined Gaussian elimination within the ring. In the 3-dimensional case 1-dimensional FFT's are performed, first local to a processor, then a new set of 1-dimensional FFT's are performed within a ring, resulting in P^2 independent tridiagonal systems, with each tridiagonal system spread across the rings. The tridiagonal systems are solved by Gaussian elimination.

Whether Gaussian elimination or cyclic reduction is preferable with respect to computational complexity for the solution of the tridiagonal systems depends on the ensemble topology, N, P, and the arithmetic rate, the communication rate, and the overhead in these operations. Gaussian elimination requires $4\frac{P^2}{N} + 2P + \alpha\sqrt{N}$ for a $\sqrt{N} \times \sqrt{N}$ mesh, substructuring with Gaussian elimination for the reduced system requires $9\frac{P^2}{N} + 4\frac{P}{\sqrt{N}} + (2+\alpha)\sqrt{N}$, and substructuring with cyclic reduction for the reduced system on a Boolean n-cube $9\frac{P^2}{N} + \frac{P}{\sqrt{N}}(3+2\alpha)log_2 N$. Precomputed coefficients are assumed for these estimates. Cyclic reduction for the reduced system becomes competitive for N approaching P on a Boolean cube configured ensemble [58,62]. On a linear array the logarithmic term premultiplied by α is replaced by a term linear in $\sqrt{N}$, as for Gaussian elimination. Which method is preferable on a linear array is critically dependent upon architectural parameters, N, and P. An accurate comparison should also account for the truncation of the reduction process for a large fraction of the systems. With respect to performance the benefit of truncating the reduction process is particularly large on linear arrays, since the largest communication expense occurs in the last few reduction steps using an *in-place* algorithm [58].

2.4.8 Iterative methods

Conjugate Gadient Methods

The conjugate gradient method [36] is a direct method for the solution of linear systems of equations. However, it is often used as an iterative method, and combined with preconditioning is an effective iterative technique, in particular for sparse systems. The conjugate gradient method solves a linear system of P equations in P steps. Each step requires $O(PZ)$ arithmetic operations for a system $Ax = y$ in which A has PZ non-zero elements. Hence, the arithmetic complexity is of the same order as for elimination methods, Given's rotations, and Householder transformations if the matrix A is dense. But, because of fill-in in those methods, the conjugate gradient method often yields a lower complexity for sparse systems, in particular if acceptable accuracy in the solution is obtained in less than P steps (possibly much fewer steps).

The minimum time per iteration is $O(log_2 P)$ because of global communication in each step. In each iteration an inner product including the entire state is computed, and used (distributed to all processors) in the computation of the new state. Pipelining of successive steps is not possible. The minimum parallel arithmetic complexity of the conjugate gradient method is $O(Plog_2 P)$, the same order as that of Householder's method. Preconditioning that would allow the iterative process to be terminated in less than $\frac{P}{log_2 P}$ steps could possibly yield a lower complexity, but the complexity of each step has to be included. With the original system matrix used as a preconditioner one iteration suffices, but the original system of equations has to be solved in that step.

So far very few studies have been carried out for parallel versions of the conjugate gradient method. Adams [?] has investigated the convergence of various preconditioners, and in particular their feasibility with respect to implementations on the Finite Element Machine. Saad and Sameh have investigated the conjugate gradient method on multiprocessors with shared global storage [112], and linear arrays [111]. The implementation of the preconditioned conjugate gradient method with various preconditioners has also been investigated by Kamath and Sameh [70]. They consider the solution of 2-dimensional elliptic partial differential equations on a ring of processors, and the solution of the 3-dimensional problem on a torus. The adaptation of the conjugate gradient method to binary tree architectures is described by Johnsson [57]. The effect of preconditioning on the computational complexity is analyzed. Van Rosendale [110] has proposed a modification of the inner product computation in which it is computed recursively. Only local computations are carried out in each step. However, global communication is still required in each step.

Asynchronous methods

In the classical iterative methods a number of matrix vector products are computed. Each such product requires global communication. In a highly concurrent system this global communication will limit the speed-up, unless several iteration steps can be pipelined. A large fraction of the processors in the ensemble are idle. So called asynchronous iterative methods, or chaotic relaxation, attempt to fully exploit the concurrency in multiprocessor systems by not enforcing global synchronization between each step of the iterative process. Chazan and Miranker [18] give necessary and sufficient conditions for convergence of chaotic relaxation applied to the solution of linear systems of equations. The results are extended by Miranker [95]. Baudet [6] gives necessary and sufficient conditions for convergence for nonlinear problems, and history dependent iterations, and some bounds on the efficiency, as well as some experimental results obtained on the C.mmp [133]. Recently, asynchronous iteration has also been studied by Lubachevsky and Mitra [92].

3 Summary

The capacity of an ensemble configuration can be measured in several different ways. One way is to measure the time required to perform arbitrary permutations. Such permutations of P elements on a binary tree of P nodes may require a time proportional to $P + O(log_2 P)$. Arbitrary permutations can be performed on a 2-dimensional mesh in time $6\sqrt{P}$ [128], on the shuffle-exchange network in time $log_2 P(log_2 P - 1)$, and on the Boolean cube in time $\frac{1}{2} log_2 P(log_2 P + 1)$ using the deterministic algorithm of Batcher, or with high probability in $clog_2 P$ time for c a small integer using a randomized algorithm. The cube connected cycles network has the same capability of performing arbitrary permutations.

Another way to measure the capability of an ensemble configuration is to determine to what extent one configuration can emulate another without a substantial increase in running time. Of the networks discussed here the, Boolean cube and the Cube Connected Cycles networks are the most powerful. The tree network is significantly less powerful in that the running time for many algorithms is higher by more than a constant factor.

The diameter of a configuration gives a lower bound for the time required for a given operation. Whether the diameter appears as an additive term or multiplicative factor in treating "large" problems on "small" ensembles depends on how communication paths with distinct origins and destinations intersect. We illustrated this point by forming a matrix transpose on a 2-dimensional mesh with end-around connections and on a Boolean cube. The transpose of a $\sqrt{N} \times \sqrt{N}$ matrix can be formed in $\frac{1}{2}\sqrt{N} - 1$ routing steps on the mesh configured ensemble, and $log_2 N$ routing steps on the Boolean cube. This difference becomes significant first for fairly large N. However, the transpose of a $P \times P$ matrix on a $\sqrt{N} \times \sqrt{N}$ mesh requires a time of at least order $\frac{O(P^2}{\sqrt{N})}$, and at most $\frac{1}{2}\lceil \frac{P}{\sqrt{N}} \rceil^2 + 1(1/2\sqrt{N} - 1)$ routing steps [50]. On the Boolean cube the transpose can be performed in $(\lceil P\ over\sqrt{N} \rceil^2 + log_2 N - 1)$ routing steps, an improvement by a factor of approximately $\sqrt{N}/4$ over the mesh.

The finite communication capability of ensemble configurations affects the performance adversely, sometimes significantly. The time for arithmetic operations decreases, but the time for communication may increase with the ensemble size. For most ensemble configurations and computations there exists an optimum size of the ensemble beyond which the performance decreases. For instance, in the case of the solution of tridiagonal systems of equations by combining Gaussian elimination and cyclic reduction the optimum sizes and minimum solution times are as follows for a few configurations: linear array $N_{opt} \approx \beta\sqrt{\frac{P}{\alpha}}$ and $T_{min} \approx \gamma\sqrt{P}$, 2-dimensional mesh $N_{opt} \approx \beta(P\ \alpha)^{2/3}$ and $T_{min} \approx \gamma P^{1/3}$, binary tree, shuffle-exchange, and Boolean cube networks $N_{opt} \approx \beta P/1 + \alpha$ and $T_{min} \approx \gamma log_2 P$, where α is the ratio between the arithmetic and communication bandwidths [58]. For band matrix solvers based on the partitioning technique the optimum number of processors configured as a linear array is of order $O(\sqrt{P/m})$ for matrices of bandwidth $2m + 1$, and $O(P/m)$ for binary tree, shuffle-exchange and Boolean cube networks. The corresponding solution times are of order $O(m^2\sqrt{Pm})$ and $O(m^3 + m^3 log_2(P/m))$, respectively [54]. For ensembles configured as Boolean cubes, band matrix solvers of complexity $O(m + m log_2 \frac{P}{m})$ can be devised [58].

With insufficient number of interconnections, or with inappropriate topology, different embedding strategies may have to be applied not only for different problems, but also for different phases of a given algorithm. Of relevance for many computations is the embedding of 1-dimensional, or multidimensional arrays. Linear arrays can be embedded in binary trees preserving proximity, but for d-dimensional arrays embedded in the leaves of the tree the average distance between nodes adjacent in the mesh is $(4 - 2^{-\lfloor log_2 n \rfloor})d$ when embedded in the tree. The maximum distance is of order $O(d log_2 n)$. Both 1-dimensional and multidimensional arrays can be embedded in Boolean cubes preserving proximity. If the number of elements in each dimension is slightly less than or equal to a power of 2, then this embedding is also efficient in terms of processor utilization. For the embedding of arbitrary meshes see [42,43]. The impact of a given embedding on performance is in some instances determined by the average distance between array nodes, whereas in others it is determined by the maximum distance.

Ensemble architecture algorithms can be obtained by first generating a computation graph from a description of the computation in a conventional mathematical notation, and then mapping this graph on to the ensemble. This mapping process has many characteristics in common with the mapping carried out in finding efficient systolic algorithms. But, there are also several aspects of the mapping of computation graphs on to ensemble architectures that do not require attention in the systolic case. One similarity is the need to treat temporal as well as spatial aspects of computations, with a nonuniform access time to different parts of the storage. Preserving locality is also important in both architectures. However, the embedding of the computation graph in an ensemble architecture often has to satisfy additional criteria compared to what is required in the systolic case in order to yield maximum processor utilization, or minimum solution time. The need for different embedding strategies during different phases of the execution of an algorithm may depend on the size of the problem relative to the size of the ensemble, as in the case of cyclic reduction.

With the additional sequencing of operations caused by mapping several nodes of a given level of the computation graph on to the same ensemble node, instead of distinct nodes as in the systolic case, independence of communication paths becomes an issue. If communication paths with distinct origins and destinations intersect at nodes only, and the processor can support concurrent communication on all its ports, then communication actions can be pipelined to a maximum extent. In effect, the ensemble is configured optimally for the desired operation. This issue was illustrated by performing a matrix transpose on a Boolean cube.

Another difference compared to algorithms of extremely fine grain is that whereas in such a case an efficient parallel algorithm may be ideal, in particular if it can be mapped on to an ensemble with only local communications without loss of efficiency, this is not necessarily true on an ensemble architecture. More operations are carried out in sequence, and the sequential operations count may be higher for an algorithm of minimum parallel complexity than for a sequential algorithm of minimum complexity. For instance, bitonic sort requires $O(N log_2^2 N)$ operations compared to $O(N log_2 N)$ operations for a good sequential sort. In the case of tridiagonal system solvers, cyclic reduction requires approximately twice the number of operations needed by Gaussian elimination. A combination of algorithms may yield a lower complexity than any single algorithm. In some instances, such as in the solution of tridiagonal systems by elimination methods, it may be possible to obtain the combined algorithm by algorithm transformation techniques. Elementary rules of algebra may be used to reduce the number of arithmetic operations carried out sequentially, as in mapping the computation of the Discrete Fourier transform on to a linear array. The result is an FFT algorithm with defined data and control structure [65]. However, the most interesting aspects of algorithm transformation techniques is that a user may not have to worry about all the minute variations of algorithms and architectural details, and that for ensemble architectures more efficient algorithms may be discovered.

In the architectural model used here it is essential that the control of execution is distributed, in order to prevent bottlenecks and avoid sources of limited scalability. All of the algorithms presented here have local control, including the routing algorithms. The architecture allows each node to execute a substantially different piece of code. However, in most of the concurrent algorithms we know there is a high degree of regularity, not only in the communication pattern, but also in the instruction streams being executed. Typically there are 3 - 4 different pieces of code. In algorithms for 2-dimensional meshes boundary nodes often perform somewhat different tasks, like computing rotation factors in the case of Given's method. In binary tree algorithms the root, the leaves, and the intermediate level nodes often have their unique pieces of code [12]. This characteristic also simplifies the problem of downloading code if the ensemble serves as an attached processor. The code can be replicated within the ensemble [90], and thereby considerably reduce the potential bottleneck caused by external input/output operations.

A large class of problems not discussed here is that of computations with data dependent control flow. For data independent computations it is possible in principle to map the computations on to the nodes in the multiprocessor system at "compile time". For simple problems mappings that are optimal with respect to some criteria, like time, can be found at a small or moderate expense. However, finding optimal mappings for most problems is, in general, an NP-complete problem. For data dependent computations good strategies for run time mappings of computations on to processors are needed. To avoid potential bottlenecks it is desirable that load balancing use only local information, and that global information is gathered through a sequence of local communications.

Acknowledgement

The author is grateful to Andrea Pappas for assistance with the illustrations. The author is also indebted to the Office of Naval Research for providing financial support under contract N00014-84-K-0043.

References

[1] Loyce Adams. *Iterative Algorithms for Large Sparse Linear Systems on Parallel Computers.* Technical Report 166027, NASA Langley Research Center, 1982.

[2] Hassan M. Ahmed, Jean-Marc Delosme, and Martin Morf. Highly concurrent computing structures for matrix arithmetic and signal processing. *Computer*, 15:65–82, January 1982.

[3] R. Aleliunas. Randomized parallel computation. In *ACM Symposium on Principles of Distributed Computing*, pages 60–72, ACM, 1982.

[4] A.V. Aho A.V., John E. Hopcroft, and Jeffrey D. Ullman. *Data Structures and Algorithms.* Addison-Wesley, 1983.

[5] Kenneth E. Batcher. Sorting networks and their applications. In *Spring Joint Computer Conference*, pages 307–314, IEEE, 1968.

[6] Gerald M. Baudet. Asynchronous iterative methods for multiprocessors. *J. ACM*, 25(2):226–244, 1978.

[7] Sandeep N. Bhatt, F.R.K. Chung, F. Tom Leighton, and Arnold L. Rosenberg. *Optimal Embeddings of Binary Trees in the Boolean Hypercube.* Technical Report YALEU/CSD/RR-, Yale University, Dept. of Computer Science, December 1985.

[8] Sandeep N. Bhatt and Ilse I.F. Ipsen. *How to Embed Trees in Hypercubes.* Technical Report YALEU/CSD/RR-443, Yale University, Dept. of Computer Science, December 1985.

[9] Sandeep N. Bhatt and Charles E. Leiserson. *Minimizing the Longest Edge in a VLSI Layout.* Technical Report MIT VLSI Memo 82-86, MIT, 1982.

[10] Smith B.J. Architecture and applications of the hep multiprocessor computer system. In *Real-Time Signal Processing IV, Proc of SPIE*, pages 241–248, 1981.

[11] Richard P. Brent and H.T. Kung. On the area of binary tree layouts. *Information Processing Letters*, 11(1):44–46, 1980.

[12] Sally A. Browning. *The Tree Machine: A Highly Concurrent Computing Environment.* Technical Report 1980:TR:3760, Computer Science, California Institute of Technology, January 1980.

[13] P. Budnik and David J. Kuck. The organisation and use of parallel memories. *IEEE Trans. Computer*, C-20:1566–1569, December 1971.

[14] Billy L. Buzbee. A fast poisson solver amenable to parallel computation. *IEEE Trans. Computers*, C-22:793–796, 1973.

[15] Billy L. Buzbee, Gene H. Golub, and C W. Nielson. On direct methods for solving poisson's equations. *SIAM J. Numer. Anal.*, 7(4):627–656, December 1970.

[16] L.E. Cannon. *A Cellular Computer to Implement the Kalman Filter Algorithm.* PhD thesis, Montana State University, 1969.

[17] Peter R. Capello and Kenneth Steiglitz. *Unifying VLSI Array Design with Linear Transformations of Space-Time.* Technical Report TRCS83-03, UC Santa Barbara, Dept of Computer Science, May 1982.

[18] D. Chazan and Willard L. Miranker. Chaotic relaxation. *Linear Algebra and its Applications*, 2:199–222, 1969.

[19] Marina C. Chen. Synthesizing systolic designs. In *2nd InternationalSymposium on VLSI Technology, Systems, And Applications*, IEEE Computer Society, 1985.

[20] W. Crowther, J. Goodhue, E. Starr, R. Thomas, W. Milliken, and T. Blackadar. Performance measurements on a 128-node butterfly parallel processor. In *Proceedings of the 1985 International Conference on Parallel Processing*, pages 531–540, IEEE Computer Society, 1985.

[21] E. Dekel, D. Nassimi, and Sartaj Sahni. Parallel matrix and graph algorithms. *SIAM J. Computing*, 10:657–673, 1981.

[22] Jean-Marc Delosme and Ilse C.P. Ipsen. An illustration of a methodology for the construction of efficient systolic architecture in vlsi. In *2nd InternationalSymposium on VLSI Technology, Systems, And Applications*, IEEE Computer Society, 1985.

[23] R.A. DeMillod, Stanley C. Eisenstat, and Richard J. Lipton. Preserving average proximity in arrays. *Communicationsof the ACM*, 21:228–231, March 1978.

[24] Jack Dongarra and S. Lennart Johnsson. Solving banded systems on a parallel processor. *Parallel Computing*, 1987. Presented at International Conference on Vector and Parallel Computing, 1986.

[25] Jack J. Dongarra and Ahmed H. Sameh. *On Some Parallel Bandded System Solvers.* Technical Report ANL/MCS-TM-27, Argonne National Laboratories, 1984.

[26] J.O. Eklundh. A fast computer method for matrix transposing. *IEEE Trans. Computers*, C-21(7):801–803, 1972.

[27] Michael J. Fischer. *Efficiency of Equivalence Algorithms*, pages 153–167. Plenum Press, 1972.

[28] Dennis Gannon and John Van Rosendale. On the impact of communication complexity in the design of parallel numerical algorithms. *IEEE Trans. Computers*, C-33(12):1180–1194, December 1984.

[29] W. Morven Gentleman. Some complexity results for matrix computations on parallel processors. *J. ACM*, 25(1):112–115, January 1978.

[30] W. Morven Gentleman and H.T. Kung. Matrix triangularization by systolic arrays. In *Real-Time Signal Processing IV, Proc. of SPIE*, pages 19–26, SPIE, 1981.

[31] Allan Gottlieb, R. Grishman, Clyde P. Kruskal, K.P. McAuliffe, L. Rudolph, and M. Snir. The nyu ultracomputer - designing an mimd shared memory parallel computer. *IEEE Trans. Computers*, C-32(2):175–189, 1983.

[32] J.W. Greene and A. El Gammal. Area and delay penalties in restructurable wafer-scale arrays. In *Third Caltech Conference on VLSI*, pages 165–184, Computer Sciences Press, 1983.

[33] Donald E. Heller and Ilse C.P. Ipsen. Systolic networks for orthogonal equivalence transformations and their applications. In P. Penfield Jr, editor, *Proceedings, Advanced Research in VLSI*, pages 113–122, Artech House, 1982.

[34] John L. Hennessey, N. Jouppi, Forrest Baskett, and J. Gill. Mips: a vlsi processor architecture. In *VLSI Systems and Computations*, pages 337–346, Computer Sciences Press, 1981.

[35] John L. Hennessey, N. Jouppi, S. Przybylski, and C. Rowen. Design of a high performance vlsi processor. In *Proc. of the Third Caltech Conference on VLSI*, pages 33–54, Computer Sciences Press, 1983.

[36] M.R. Hestenes and E. Stiefel. Methods of conjugate gradient for solution of linear systems. *J. Res. Nat. Bur. Standards*, 49:409–436, 1952.

[37] W. Daniel Hillis. *The Connection Machine.* MIT Press, 1985.

[38] W. Daniel Hillis. *The Connection Machine.* Technical Report Memo 646, MIT Artificial Intelligence Laboratory, 1981.

[39] W. Daniel Hillis and Guy L. Steel. Data parallel algorithms. *Communications of the CACM*, 29:1170–1183, December 1986.

[40] Daniel S. Hirschberg. Fast parallel sorting algorithms. *Communications of the ACM*, 21(8):657–661, 1978.

[41] Ching-Tien Ho and S. Lennart Johnsson. *Matrix Transposition on Boolean n-cube Configured Ensemble Architectures.* Technical Report YALEU/CSD/RR-494, Yale University, Dept. of Computer Science, September 1986.

[42] Ching-Tien Ho and S. Lennart Johnsson. On the embedding of arbitrary meshes in boolean cubes with expansion two dilation two. In *Int. Conf.on Parallel Processing*, IEEE Computer Society, 1987.

[43] Ching-Tien Ho and S. Lennart Johnsson. *On the Embedding of Meshes in Boolean Cubes.* Technical Report YALEU/CSD/RR-, Yale University, Dept. of Computer Science, In preparation 1986.

[44] Ching-Tien Ho and S. Lennart Johnsson. *Spanning Graphs for Optimum Broadcasting and Personalized Communication in Hypercubes.* Technical Report YALEU/CSD/RR-500, Yale University, Dept. of Computer Science, November 1986.

[45] Roger W. Hockney. A fast direct solution of poisson's equation using fourier analysis. *J. ACM*, 12:95–113, 1965.

[46] Roger W. Hockney. The potential calculation and some applications. *Methods Comput. Phys.*, 9:135–211, 1970.

[47] Roger W. Hockney and C.R. Jesshope. *Parallel Computers.* Adam Hilger, 1981.

[48] Briggs F.A. Hwang K., editor. *Computer Architecture and Parallel Processing.* McGraw-Hill, 1984.

[49] S. Lennart Johnsson. Combining parallel and sequential sorting on a boolean n-cube. In *International Conference on Parallel Processing*, pages 444–448, IEEE Computer Society, 1984. Presented at the 1984 Conf. on Vector and Parallel Processors in Computational Science II.

[50] S. Lennart Johnsson. Communication efficient basic linear algebra computations on hypercube architectures. *Journal of Parallel and Distributed Computing*, 4(2):133–172, April 1987. Report YALEU/CSD/RR-361, January 1985, Dept. of Computer Science, Yale University.

[51] S. Lennart Johnsson. A computational array for the qr-method. In Jr. P. Penfield, editor, *Proc., Conf. on Advanced Research in VLSI*, pages 123–129, Artech House, January 1982.

[52] S. Lennart Johnsson. *Computational Arrays for Band Matrix Equations.* Technical Report 4287:TR:81, Computer Science, California Institute of Technology, May 1981.

[53] S. Lennart Johnsson. *Data Permutations and Basic Linear Algebra Computations on Ensemble Architectures.* Technical Report YALEU/CSD/RR-367, Yale University, Dept. of Computer Science, February 1985.

[54] S. Lennart Johnsson. Dense matrix operations on a torus and a boolean cube. In *The National Computer Conference*, July 1985.

[55] S. Lennart Johnsson. *Fast Banded Systems Solvers for Ensemble Architectures.* Technical Report YALEU/CSD/RR-379, Department of Computer Science, Yale University, March 1985.

[56] S. Lennart Johnsson. Fast pde solvers on fine and medium grain architectures. In *Int. Assoc. for Mathematics and Computers in Simulation*, page , IMACS, 1987.

[57] S. Lennart Johnsson. Highly concurrent algorithms for solving linear systems of equations. In *Elliptic Problem Solving II*, Academic Press, 1983.

[58] S. Lennart Johnsson. *Odd-Even Cyclic Reduction on Ensemble Architectures and the Solution Tridiagonal Systems of Equations.* Technical Report YALE/CSD/RR-339, Department of Computer Science, Yale University, October 1984.

[59] S. Lennart Johnsson. Pipelined linear equation solvers and vlsi. In *Microelectronics '82*, pages 42–46, Institution of Electrical Engineers, Australia, May 1982.

[60] S. Lennart Johnsson. Solving narrow banded systems on ensemble architectures. *ACM TOMS*, 11(3):271–288, November 1985. Also available as Report YALEU/CSD/RR-418, November 1984.

[61] S. Lennart Johnsson. *Solving Narrow Banded Systems on Ensemble Architectures.* Technical Report YALEU/CSD/RR-343, Dept. of Computer Science, Yale University, November 1984.

[62] S. Lennart Johnsson. Solving tridiagonal systems on ensemble architectures. *SIAM J. Sci. Stat. Comp.*, 8(3):354–392, May 1987. Report YALEU/CSD/RR-436, November 1985.

[63] S. Lennart Johnsson. Vlsi algorithms for doolittle's, crout's and cholesky's methods. In *International Conference on Circuits and Computers 1982, ICCC82*, pages 372–377, IEEE, Computer Society, September 1982.

[64] S. Lennart Johnsson and Danny Cohen. An algebraic description of array implementations of fft algorithms. In *20th Allerton Conference on Communication, Control, and Computing*, Electrical Engineering, University of Illinois, Urbana/Champaign, 1982.

[65] S. Lennart Johnsson and Danny Cohen. *Mathematical Approach to Computational Networks for the Discrete Fourier Transform.* Technical Report, Department of Computer Science, Yale University, 1984.

[66] S. Lennart Johnsson and Ching-Tien Ho. Matrix multiplication on boolean cubes using generic communication primitives. In *Parallel Processing and Medium Scale Multiprocessors*, SIAM, 1987. YALEU/CSD/RR-530.

[67] S. Lennart Johnsson, Ching-Tien Ho, and Faisal Saied. *Multiple tridiagonal systems, the Alternating Direction Method, and Boolean cube configured multiprocessors.* Technical Report , Yale University, In preparation 1987.

[68] S. Lennart Johnsson, Uri Weiser, Danny Cohen, and Al Davis. Towards a formal treatment of vlsi arrays. In *Proceedings of the Second Caltech Conference on VLSI*, pages 375 – 398, Caltech Computer Science Department, January 1981.

[69] Hwang K., editor. *Supercomputers: Design and Applications.* IEEE Computer Society, 1984.

[70] C. Kamath and Ahmed H. Sameh. *The Preconditioned Conjugate Gradient Method on a Multiprocessor.* Technical Report ANL/MCS-TM-28, Argonne National Laboratories, Mathematics and Computer Science Division, 1984.

[71] M.G.H. Katevenis. *Reduced Instruction Set Computer Architectures for VLSI.* The MIT Press, 1985.

[72] D. Kershaw. *Solution of Single Tridiagonal Linear Systems and the Vectorization of the ICCG Algorithm on the CRAY-1*, pages 85–92. Academic Press, 1982.

[73] S.C. Knauer, J.H. O'Neill, and A. Huang. *Self-routing Switching Network*, pages 424–448. Addison-Wesley, 1985.

[74] Donald E. Knuth. *The Art of Computer Programming, Vol. 3: Sorting and Searching.* Addison-Wesley, 1973.

[75] P.M. Kogge and Harold S. Stone. A parallel algorithm for the efficient solution of a general class of recurrence equations. *IEEE Trans. Computers*, C-22(8):786–792, 1973.

[76] David J. Kuck. *The Structure of Computers and Computations.* John Wiley, 1978.

[77] David J. Kuck. A survey of parallel machine organization and programming. *ACM Computing Surveys*, 9(1):29–59, 1977.

[78] David J. Kuck, Duncan H. Lawrie, R. Cytron, Ahmed Sameh, and Daniel D. Gajski. *The Architecture and the Programming of the Cedar System.* Technical Report, Laboratory for Advanced Supercomputers, Dept. of Computer Science, University of Illinois, August 1983.

[79] M. Kumar and Daniel S. Hirschberg. An efficient implementation of batcher's bitonic odd-even merge algorithm and its application in parallel sorting schemes. *IEEE Trans. Computers*, C-32(3):254–264, 1983.

[80] H.T. Kung and Charles E. Leiserson. *Algorithms for VLSI Processor Arrays*, pages 271–292. Addison-Wesley, 1980.

[81] Snyder L. Introduction to the configurable highly parallel computer. *Computer*, 15(1):47–56, 1982.

[82] Duncan H. Lawrie. Access and alignment of data in an array processor. *IEEE Trans. on Computers*, C-24(12):99–109, 1975.

[83] Duncan H. Lawrie and Ahmed H. Sameh. The computational and communication complexity of a parallel banded system solver. *ACM TOMS*, 10(2):185–195, June 1984.

[84] Duncan H. Lawrie and C.R. Vora. The prime memory system for array access. *IEEE Trans. Computer*, C-31:1435–442, May 1982.

[85] F. Tom Leighton. *Complexity Issues in VLSI: Optimal Layouts for the Shuffle-Exchange Graph and Other Networks.* MIT Press, 1983.

[86] F. Tom Leighton and Charles E. Leiserson. Wafer-scale integration of systolic arrays. *IEEE Trans. Comp.*, C-34(5):448–461, May 1985.

[87] Charles E. Leiserson. *Area-Efficient VLSI Computation.* MIT Press, 1982.

[88] Charles E. Leiserson. Fat-trees: universal networks for hardware-efficient supercomputing. *IEEE Trans. Computers,* C-34:892–901, October 1985.

[89] Li.-J. Li and Benjamin W. Wah. The design of optimal systolic arrays. *IEEE Trans. Computers,* C-34:66–77, 1985.

[90] Peggy Li and Lennart Johnsson. The tree machine: an evaluation of program loading strategies. In *1983 International Conference on Parallel Processing,* pages 202 – 205, IEEE Computer Society, August 1983.

[91] Bjorn Lisper. *Description and Synthesis of Systolic Arrays.* Technical Report TRITA-NA-8318, The Royal Institute of Technology, Dept. of Numerical Analysis and Computing Sciences, 1983.

[92] Boris Lubachevsky and Debasis Mitra. *A Chaotic, Asynchronous Algorithm for Computing the Fixed Point of a Nonnegative Matrix of Unit Spectral Radius.* Technical Report, AT&T Bell Laboratories, 1984.

[93] Christoffer Lutz, Steve Rabin, Charles L. Seitz, and Donald Speck. Design of the mosaic element. In *Proceedings, Conf. on Advanced research in VLSI,* pages 1–10, Artech House, 1984.

[94] Carver A. Mead and Lynn Conway. *Introduction to VLSI Systems.* Addison-Wesley, 1980.

[95] Willard L. Miranker. Hierarchical relaxation. *Computing,* 23:267–285, 1979.

[96] Willard L. Miranker and Andrew Winkler. Spacetime representations of computational structures. *Computing,* 32(2):93–114, 1984.

[97] Donald I. Moldovan. On the design of algorithms for vlsi systolic arrays. *Proc. IEEE,* 71(1):113–120, 1983.

[98] D. Nassimi and Sartaj Sahni. Bitonic sort on a mesh-connected parallel computer. *IEEE Trans. Computers,* C-27(1):2 – 7, 1979.

[99] Susan T. O'Donnel, P. Geiger, and Martin H. Schultz. *Solving the Poisson Equation on the FPS-164.* Technical Report YALEU/DCS/RR-293, Research Center for Scientific Computing, Dept. of Computer Science, Yale University, November 1983.

[100] M.S. Paterson, W.L. Ruzzo, and Larry Snyder. Bounds on minimax edge length for complete binary trees. In *Proc. of the 13th Annual Symposium on the Theory of Computing,* pages 293–299, ACM, 1981.

[101] Gregory F. Pfister, W.C. Brantley, D.A. George, S.L. Harvey, W.J. Kleinfelder, K.P. McAuliffe, E.A. Melton, V.A. Norton, and J. Weiss. The ibm research parallel processor prototype (rp3); introduction and architecture. In *Proceedings of the 1985 International Conference on Parallel Processing,* pages 764–771, IEEE Computer Society, 1985.

[102] Franco P. Preparata and J.E. Vuillemin. The cube connected cycles: a versatile network for parallel computation. In *Proc. Twentieth Annual IEEE Symposium on Foundations of Computer Science,* pages 140–147, 1979.

[103] P. Quinton. Automatic synthesis of systolic arrays from uniform recurrent equations. In *Proc. 11th Annual Symposium on Computer Architecture,* pages 208–214, IEEE Computer Society, 1984.

[104] L.R. Rabiner and B. Gold. *Theory and Application of Digital Signal Processing.* Prentice-Hall, 1975.

[105] Abhiram Ranade. Interconnection networks and parallel memory organization for array processing. In *1985 International Conference on Parallel Processing*, IEEE Computer Society, 1985.

[106] Abhiram Ranade and S. Lennart Johnsson. The communication efficiency of meshes, boolean cubes, and cube connected cycles for wafer scale integration. In *Int. Conf. on Parallel Processing*, page , IEEE Computer Society, 1987.

[107] E M. Reingold, J Nievergelt, and N Deo. *Combinatorial Algorithms.* Prentice Hall, 1977.

[108] E. Reiter and Gary Rodrigue. An incomplete cholesky factorization by a matrix partitioning algorithm. In *Elliptic Problem Solvers II*, pages 161–174, Academic Press, 1983.

[109] Arnold L. Rosenberg and Larry Snyder. Bounds on the costs of data encodings. *Mathematical Systems Theory*, 12:9–39, 1978.

[110] John Van Rosendale. Minimizing inner product dependencies in conjugate gradient iteration. In *Proc. of the 1983 International Conference on Parallel Processing*, pages 44–46, IEEE Computer Society, 1983.

[111] Yousef Saad. *Practical use of Polynomial Preconditionings for the Conjugate Gradient Method.* Technical Report YALEU/DCS/RR-282, Dept. of Computer Science, 1983.

[112] Yousef Saad and Ahmed H. Sameh. Iterative methods for the solution of elliptic differential equations on multiprocessors. In *Proc. of the CONPAR 81 Conference*, pages 395–411, Springer Verlag, 1981.

[113] Faisal Saied, Ching-Tien Ho, S. Lennart Johnsson, and Martin H. Schultz. Solving schroedinger's equation on the intel ipsc by the alternating direction method. In *Hypercube Multiprocessors 1987*, page , SIAM, September 1986. Tech. report YALEU/CSD/RR-502.

[114] Ahmed H. Sameh. A fast poisson solver for multiprocessors. In *Elliptic Problem Solvers II*, pages 175–186, Academic Press, 1984.

[115] Ahmed H. Sameh. Numerical parallel algorithms - a survey. In *High Speed Computer and Algorithm Organization*, pages 207–228, Academic Press, 1977.

[116] Ahmed H. Sameh and David J. Kuck. On stable parallel linear system solvers. *J. ACM*, 25(1):81–91, January 1978.

[117] J.T. Schwartz. Ultracomputers. *ACM Trans. on Programming Languages and Systems*, 2:484–521, 1980.

[118] Charles L. Seitz. Ensemble architectures for vlsi – a survey and taxonomy. In P. Penfield Jr., editor, *1982 Conf on Advanced Research in VLSI*, pages 130 – 135, Artech House, January 1982.

[119] Charles L. Seitz. Experiments with vlsi ensemble machines. *J. VLSI Comput. Syst.*, 1(3), 1984.

[120] M.C. Sejnowski, E.T. Upchurch, R.N. Kapur, D.P.S. Charlu, and G.J. Lipovski. An overview of the texas reconfigurable array computer. In *Proceedings, National Computer Conference*, pages 631–641, IEEE, 1980.

[121] M. Sekanina. On an ordering of the set of vertices of a connected graph. *Publ. of the Faculty of the Sciences of the Univ. of Brno*, 412():137–142, 1960.

[122] J. Stevens. *A Fast Fourier Transform Subroutine for the Illiac IV.* Technical Report, Center for Advanced Computation, Univ. of Illinois, 1971.

[123] Harold S. Stone. Parallel processing with the perfect shuffle. *IEEE Trans. Computers*, C-20:153–161, 1971.

[124] Paul N. Swartztrauber. The methods of cyclic reduction, fourier analysis, and the facr algorithm for the discrete solution of poisson's equation on a rectangle. *SIAM Review*, 19:490–501, 1977.

[125] Clive Temperton. Direct methods for the solution of the discrete poisson equation: some comparisons. *J. of Computational Physics*, 31:1–20, 1979.

[126] Clive Temperton. On the facr(l) algorithm for the discrete poisson equation. *J. of Computational Physics*, 34:314–329, 1980.

[127] C.D. Thompson. *Fourier Transforms in VLSI.* Technical Report UCB/ERL/ M80/51, Electronic Research Laboratory, UC Berkeley, 1980.

[128] C.D. Thompson and H.T. Kung. Sorting on a mesh-connected parallel computer. *CACM*, 20(4):263–271, 1977.

[129] Jeffrey D. Ullman. *Computational Aspects of VLSI.* Computer Sciences Press, 1984.

[130] E. Upfal. Efficient schemes for parallel computation. In *ACM Symposium on Principles of Distributed Computing*, pages 55–59, ACM, 1982.

[131] Leslie Valiant and G.J. Brebner. Universal schemes for parallel communication. In *Proc. of the 13th ACM Symposium on the Theory of Computation*, pages 263–277, ACM, 1981.

[132] A.Y. Wu. Embedding of tree networks in hypercubes. *Journal of Parallel and Distributed Computing*, 2(3):238–249, 1985.

[133] W.A. Wulf and C.G. Bell. C.mmp - a multi-mini-processor. In *AFIPS 72 FJCC*, pages 765–777, 1972.

PROBLEM DECOMPOSITION AND COMMUNICATION TRADEOFFS IN A SHARED-MEMORY MULTIPROCESSOR

Thomas J. LeBlanc
Computer Science Department
University of Rochester
Rochester, New York 14627

Abstract

Conceptually, a tightly-coupled, shared-memory multiprocessor, such as the BBN Butterfly Parallel Processor, can support a model of computation based on either shared memory or message passing. The choice of model affects the decomposition of the problem into parallel processes and the resultant granularity of communication. An architecture, such as the Butterfly, that supports both models offers the programmer a wide range of choices for problem decomposition, each with different performance attributes. This paper describes a series of experiments using Gaussian elimination to evaluate the tradeoffs between shared memory and message passing, over a range of decomposition strategies. Several different implementations are described and their performance is compared. We conclude that the performance of an application depends not only on the efficiency of the underlying communication, but also on the extent to which the underlying model of computation encourages or discourages communication.

1. Introduction

Programming a multiprocessor (or any other parallel computer) involves tradeoffs between parallelism and communication. In order to achieve the best performance, the programmer must maximize parallelism and minimize communication. These goals are in conflict since increased parallelism usually implies increased communication. Although the intrinsic properties of the application limit the amount of parallelism available, communication costs so frequently dominate that the inherent limit to parallelism is rarely reached. At one extreme, we can minimize communication by using single processor systems, but these systems don't admit parallelism. At the other extreme, local-area networks offer a large degree of parallelism, but communication in these networks is relatively expensive. In such an environment, the decomposition of a problem into processes is determined more by communication requirements than intrinsic parallelism. Recently available commercial multiprocessors, such as Sequent's Balance 8000 [7] and Alliant's FX Series [1], offer very efficient communication, but limit the parallelism to some small constant factor less than 16. One of the few commercially available multiprocessors that does not put such a low limit on parallelism and yet still provides efficient communication between processors is the BBN Butterfly Parallel Processor™.

The Butterfly hardware architecture supports message passing between processors using an FFT network of 4x4 switch elements. The memory architecture, implemented by the operating system in conjunction with a micro-coded co-processor, provides the illusion of shared memory. Software packages have been developed that support the shared-memory illusion and the message-passing model. This paper describes a series of experiments designed to explore the tradeoffs between problem decomposition strategies and communication costs on a shared-memory multiprocessor such as the Butterfly.

Both shared-memory and message-passing models have been advocated for parallel computation. Shared memory is usually found in tightly-coupled architectures that support remote memory operations in hardware or firmware. Efficiency is the primary motivation for allowing shared memory at the user level. Every process can retrieve the data it requires using hardware primitives; no costly system software need be involved. Often, this approach ignores the problem of synchronization, which can dominate the cost of remote references. Shared memory is useful in applications that communicate values which must be interpreted in the context of an environment (*e.g.*, pointers). Since shared memory can be used to store an inherited context for a computation, this model makes it practical to consider small-grain decomposition strategies.

Message passing is typically employed in loosely-coupled systems (*e.g.*, local-area networks). Since the performance characteristics and error rates of remote communication differ significantly with those of local communication, it is natural to use two different primitives for remote and local communication. Shared memory supported by hardware is used for local references, while non-local communication uses message passing, usually implemented by system software. Message passing has two advantages over shared memory. First, an exchange of messages is a more abstract form of communication than reading a memory location. Message passing subsumes communication, buffering, and synchronization. Second, processes in a message-passing system cannot directly affect one another. This *object-oriented* approach has proven quite popular for distributed systems composed of many processes. Errors can be isolated by monitoring the stream of messages produced by the various processes. A corresponding disadvantage of message passing is that it imposes a *value-oriented* semantics. Processes may only communicate values, which may require the exchange of an environment in which to interpret the value (*i.e.*, a pointer value is represented by the object to which it points). The inability to inherit a context easily means that the message-passing model encourages large-grain decomposition strategies.

Early work on the Butterfly, the Voice Funnel application in particular, used message passing. More recently Butterfly applications, including finite element analysis and computer vision algorithms, have assumed the shared-memory model of computation. The only programming environment available from BBN on the Butterfly that masks the low-level details from the programmer is the Uniform System package, which implements a shared-memory model. Programmers find it easier to use the Uniform System than to build the program from scratch, *regardless* of how well the application fits the shared-memory model. In order to provide the Butterfly programmer with implementation alternatives, we have constructed a library package based on message passing, called SMP [4].

We conducted a series of experiments, using Gaussian elimination as a sample application, to evaluate the tradeoffs inherent in both the hardware and software architectures. The goal of the experiments was to explore the tradeoffs between different decomposition and communication strategies using the shared-memory model, as implemented by the Uniform System package, and a simple message-passing model, as implemented by SMP. In this paper, we present the results of these experiments.

In Section 2 we provide an overview of the Butterfly Parallel Processor hardware architecture and software environments. Section 3 describes the sample application we chose for the experiments and the decomposition strategies, with related communication costs, we studied. In Section 4, we describe several different implementations of the sample application using both shared memory and message passing for communication. Section 5 compares the performance of the various implementations. Conclusions based on our experiences are presented in Section 6.

2. Butterfly Hardware and Software Overview

The BBN Butterfly Parallel Processor™ consists of up to 256 processing nodes connected by a switching network. Each processor is an 8 MHz MC68000 with 24 bit virtual addresses. A 2901-based bit-slice co-processor interprets every memory reference issued by the 68000 and is used to communicate with other nodes across the switching network. All of the memory in the system resides on individual nodes, but any processor can address any memory through the switch. A remote memory reference (read) takes about 4 *us.*, roughly 5 times as long as a local reference.

Each switch node in the switching network is a 4-input, 4-output crossbar switch with a bandwidth of 32 megabits/sec. An N processor system uses $(N \log_4 N)/4$ switches arranged in $\log_4 N$ columns. A 128-node Butterfly contains 256 switch nodes; the extra switch capacity is used to provide an alternate communication path between all processors for reliability and improved efficiency. The alternate paths can be enabled or disabled by the user, making it possible to indirectly measure the effect of switch contention.

The Butterfly is programmed using the C programming language augmented with system calls to Chrysalis, the Butterfly operating system [5]. Chrysalis is a protected subroutine library that implements various operations for process management, interprocess communication, and memory management. The most common operations are implemented by microcode. Chrysalis is too low-level for application programmers, however, the primitive operations provided by Chrysalis offer a general framework upon which efficient high-level communication protocols and software systems can be built. Two such systems are the Uniform System package from BBN [2] and SMP [4], a message-passing library implemented at the University of Rochester.

2.1. The Uniform System

The Uniform System (US) supports a user-level view of the architecture consisting of lightweight tasks and a globally-shared memory. The shared memory is implemented by the virtual address space of the Butterfly; the lightweight tasks are implemented by manager processes, one on each node. To reduce memory contention (*i.e.*, hot-spots), US encourages the scattering of data uniformly throughout the machine. This may even include putting data in memory associated with a processor that is not otherwise involved in the computation. To reduce process management overhead, only one manager process is allocated per node; all tasks that run on a node are executed by that node's manager process. Since each task inherits the globally-shared memory upon creation, US supports a very small task granularity.

The US library consists of calls to create a globally-shared memory, spread data throughout the shared memory, and create tasks that operate on the shared memory. The shared memory is accessible to all US tasks. Pointers to storage within the shared memory may be shared between processors. During initialization, US creates the manager process for each processor, which is responsible for allocating the processor to a series of tasks. Usually, a task is some small procedure to be applied to a subset of the shared memory. A task, therefore, can be represented as simply an index, or a range of indices, into the shared memory and an operation to be performed on that memory. Atomic operations in microcode are used to efficiently allocate tasks to processors. Tasks may then use the shared memory for interprocess communication. Busy-wait locks are provided for synchronization; other synchronization primitives are also available directly from Chrysalis.

One of the main purposes of US is to hide the details of the architecture. US programs can be written independent of the number of processors, since the user is not aware of how tasks are allocated to processors. In addition, US supports two classes of memory: private memory for each process and the tasks it represents, and globally-shared memory. However, even though all of the memory in the system resides on individual nodes, the distinction between local and remote memory, as defined by the architecture, is not applied to the globally-shared memory. This design has important ramifications on the communication costs we will consider.

2.2. SMP

SMP (Structured Message Passing) views the Butterfly as a fast message-passing machine. Although the memory architecture is used to implement message buffers, user processes cannot share memory. SMP is similar in flavor and scope to the Uniform System. SMP provides Butterfly programmers with a model of parallel programs that consists of: (1) *process families*, whose members are created and destroyed together, (2) interprocess communication, within a family, based on *asynchronous message passing* (send/receive) according to a fixed communication topology, and (3) *a dynamic hierarchy* of such process families. SMP process families and hierarchies add structure to the basic process model of Chrysalis. The advantages of message passing include increased autonomy for processes, increased efficiency by exploiting locality of data, avoidance of explicit synchronization for data access, and improved protection between processes.

The SMP communication primitives are **Send** and **Receive**. Communication is asynchronous between a sender and receiver. **Send** may specify multiple destinations, providing a form of multicast communication. It is nonblocking in that the sender may continue to execute before all destinations receive the message. **Receive** will block until a message from the source arrives. A "wildcard" value is available so that messages from any source may be received. Although there are additional library calls to create processes, there are no other communication or synchronization primitives in SMP. In particular, SMP does not support any form of shared memory.

SMP processes are implemented by Chrysalis processes. These heavyweight processes are expensive to create, but have an efficient context switch mechanism. Communication between SMP processes uses a single primitive that subsumes communication and synchronization. As a result, communication is more expensive than reading a remote memory location. Therefore, SMP encourages the use of a relatively small number of SMP processes consisting of significant computation segments that communicate infrequently. As with US, SMP does not distinguish between local and remote memory, since only local accesses are allowed. However, the distinction between cheap communication (local accesses) and expensive communication (messages) is explicit in SMP.

3. Sample Application: Gaussian Elimination

We chose Gaussian elimination (without pivoting) as the sample application for our experiments. The primary reason for this choice was that several experiments using shared memory to implement Gaussian elimination had already been performed at BBN Laboratories [3,8]. Our experiments were designed for comparison with their results. However, we were also interested in Gaussian elimination as a representative of a large class of algorithms in finite element analysis. Although it is fairly easy to develop parallel variants of Gaussian elimination, the problem does have interesting synchronization constraints and several different decomposition strategies are possible. Our experiments were designed to explore the decomposition and communication tradeoffs in Gaussian elimination; it was not our intention to explore efficient algorithms for solving linear systems, such as those found in [6].

In solving a set of linear equations using Gaussian elimination, the coefficient matrix M is diagonalized, producing a modified vector of unknowns, and the unknowns are determined using back-substitution. (Since back-substitution is a small percentage of the total time required to solve the equations, it is not performed in any of the experiments.) To eliminate an entry M[i,j], we replace row M[i] with M[i] - (M[j] * M[i,j]/M[j,j]), where M[j] is known as the pivot row. However, this operation cannot be performed until row M[j] has stabilized, *i.e.*, M[j,k] = 0, $\forall$ k < j. In addition, all previous entries in row i must already be eliminated, *i.e.*, M[i,k] = 0, $\forall$ k < j. These two synchronization constraints limit the amount of parallelism that we can expect to achieve.

Floating point arithmetic is required to solve a set of linear equations. Unfortunately, our Butterfly does not have floating point hardware. (A hardware floating point arithmetic unit for the Butterfly is now available from BBN as a hardware upgrade.) All floating point arithmetic is performed by costly subroutines. This makes it difficult to analyze the communication costs of an application because the execution time is dominated by floating point software. To alleviate this problem, our experiments were performed with *simulated floating point* arithmetic, using the same approach as that used in the earlier experiments [3]. All floating point variables were replaced with integer variables, and addition and subtraction was used in place of multiplication and division. Such a computation does not give the correct answers, but does accurately simulate the performance of the computation on a Butterfly with floating point hardware. Each experiment was performed using software floating point to ensure the correctness of the results, but all reported performance figures refer to the results of experiments using simulated floating point. By using simulated floating point arithmetic, we not only can make predictions about the performance of the Butterfly with floating point hardware, we also reduce the execution time of the computation and thereby increase the significance of communication costs.

3.1. Decomposition Strategies

In Gaussian elimination there are several boundaries that can be drawn for process decomposition. The problem consists of matrix elements, rows, and columns. The synchronization constraints effectively preclude delineating processes using columns since the processing of multiple columns within the same row cannot occur in parallel. Diagonalization of a single row is a reasonable unit of parallelism, since every row can eliminate the first column immediately, and subsequent columns as rows are diagonalized. Since the synchronization constraints are expressed in terms of individual elements, a finer-grain allocation is also possible, wherein each element to be eliminated is represented by a process. In addition, realizing that the parallelism of any decomposition into processes is limited by the number of processors, a coarse-grain decomposition, wherein each processor represents a process, is also possible.

We will consider three decomposition strategies: small-grain decomposition, where each element to be eliminated is a process, medium-grain decomposition, where each row to be diagonalized is a process, and large-grain decomposition, where each processor is a process and contains a subset of the rows to be diagonalized. If we assume P processors and an NxN matrix, then the small-grain decomposition has $(N^2-N)/2$ processes, the medium-grain decomposition has N processes, and the large-grain decomposition has P processes.

3.2. Communication Strategies

Within a particular decomposition strategy, we would like to minimize the time required for communication between processes. The overall cost of communication depends on the type and amount of communication. Access to shared memory is supported by the architecture and is very efficient, particularly when the granularity of communication is large enough to amortize the setup costs. (For the Butterfly, block transfers can be used profitably for any transfer larger than 4 bytes.) Message-based communication is implemented by software and is significantly more expensive. The amount of communication depends for the most part on the amount of state local to a process. Small-grain (lightweight) processes require more communication than large-grain (heavyweight) processes. In each case, we can reduce the amount of communication by transmitting only the nonzero entries in a row.

In Gaussian elimination, the basic computation step is the elimination of a single entry within a row. To perform this operation, both the row containing the entry and the pivot row are needed. In the small-grain decomposition, where each element to be eliminated is a process, neither row can be statically allocated in the state of a process. The pivot row values are not determined until the row is diagonalized and the value of the individual entry to be eliminated is not known until all preceding entries in that row are eliminated. Therefore, each of $(N^2\text{-}N)/2$ processes must copy two rows into their local state to perform their computation. This results in a total of $N^2\text{-}N$ copy operations.

In the medium-grain decomposition, each process contains a row to be diagonalized as part of its local state. Any changes to the row that occur during the elimination of some element are naturally inherited by the operations that eliminate succeeding elements. Only the corresponding pivot rows must be copied into the local state. Overall, a pivot row must be copied for each of the $(N^2\text{-}N)/2$ elements to be eliminated, resulting in $(N^2\text{-}N)/2$ copy operations.

In the large-grain decomposition, each process contains a set of rows to be diagonalized. As with the medium-grain decomposition, each pivot row must be copied into the local state, however, a single copy of a pivot row can be used to eliminate an entire column of the subset of the problem matrix stored in local state. In addition, any local rows that later act as pivot rows do not need to be copied into the local state. Therefore, each of N-1 pivot rows must be copied into P-1 processors, requiring (P-1)(N-1) copy operations.

4. Implementation Descriptions

All three decomposition granularities were implemented. The small-grain and medium-grain decompositions were implemented using the Uniform System and shared memory. The large-grain decomposition was implemented using SMP and message passing.

4.1. Uniform System Implementation

Several experiments had previously been performed at BBN Laboratories [3,8] to see how well applications that use the Uniform System, including Gaussian elimination, perform on large Butterfly configurations. All of our Uniform System implementations are based on adaptations of the program developed at BBN.

In all US experiments, the problem matrix is uniformly distributed throughout the globally-shared memory. In the small-grain decomposition case, a task is created to eliminate a single entry in the matrix, M[i,j]. Both row M[i] and row M[j] are transferred to local memory from the globally-shared memory before performing the computation, thereby avoiding a remote reference for each individual entry in a row. (Block transfers amortize overhead associated with remote references and can significantly improve performance.) Since M[i] is modified by the operation and must be used by other tasks, an additional transfer of the modified row back to the global memory location for M[i] is also necessary. The total number of copy operations is $3(N^2-N)/2$. (Note: the initialization step necessary to load the problem matrix into the globally-shared memory is not included, either in the analysis of copy operations nor in the empirical data.)

In the medium-grain decomposition, a task is created to eliminate entries in an entire row, M[i]. Row M[i] and all rows M[j], $j < i$, must be transferred to local memory from the globally-shared memory. Since each element to be eliminated requires a pivot row to be copied and each row must be copied into the local memory of the task associated with that row, the total number of copy operations is $(N^2-N)/2 + N$.

4.2. SMP Implementation

In the SMP implementation of Gaussian elimination, a coordinator process is responsible for creating worker processes on different processors. All workers initialize the local message handler, which requires global synchronization, and then create a local partition of the problem. When N processors are used, each processor P is assigned the task of initializing and then diagonalizing all rows R for which R mod N = P. To do so, each processor must request each row of the problem matrix in sequence and use it to eliminate entries in the corresponding column of the local partition. A percentage of these requests, 1/P, will be satisfied locally and, therefore, do not require communication. Local rows are broadcast to the other processors when they have been diagonalized. When all local rows are completely diagonalized, the worker process signals the coordinator, which is responsible for termination of the computation.

The total number of copy operations is P(N-1). Each of N-1 pivot rows is copied into P-1 processors. An additional copy operation for each pivot row is necessary to buffer the message locally before it is broadcast to the other processors. Since each process initializes some subset of the problem matrix in its local memory, there is no need to communicate with a global copy of the problem matrix.

4.3. Caching Strategies

Interprocess communication costs are not determined solely by the hardware organization. Although there is a clear need for communication between different processors, there is also a need for communication between different processes. In fact, the communication required between software boundaries can dominate communication between hardware boundaries, depending on the problem decomposition used. For example, in the small-grain decomposition using the Uniform System, fully one third of the copy operations are used to move data between the local memory (where it cannot be used directly by some other task) and the globally-shared memory. $3N^2/2$ copy operations are required by the US implementation, but only $2N^2/2$ copy operations are required by the decomposition. To reduce the overall amount of communication in an application, caching strategies can be used. Such strategies can be particularly effective when the data maintained in the cache changes infrequently.

In Gaussian elimination, once a row has been completely diagonalized, it will stay unchanged for the remainder of the computation. One possible strategy to reduce communication costs is to cache pivot rows. The large-grain decomposition does this naturally, since each pivot row is, in effect, cached on each processor and is used to eliminate an entire column in some subset of rows. The medium-grain and small-grain decompositions do not have a notion of processor, so much of the communication in these cases is required by the logical structure of the program, not the physical allocation of the data. Since US provides process-local data, we can cache the pivot rows on a process basis. For the medium-grain decomposition, this reduces the number of copy operations from $(N^2-N)/2 + N$ copy operations, where we copy the pivot row each time an element is eliminated, to $P(N-1) + N$ copy operations, where each pivot row is copied into each other processor. Caching the pivot rows in the small-grain decomposition reduces the number of copy operations from $3(N^2-N)/2$ to $N^2-N + P(N-1)$. These results are summarized in Table 1, which shows the number of copy operations for a problem matrix of size 800x800 for each decomposition strategy, including caching.

In the small-grain and large-grain decompositions, the cache requires little space. In the large-grain decomposition, each pivot row is used to eliminate a column in the local subset of rows before another pivot row is received. Once a pivot row has been used to eliminate a column of entries, there is no need to continue to store that pivot row. Therefore, the implicit caching of pivot rows takes little space. The same is true for the small-grain decomposition implemented by US. US makes it possible to order the allocation of tasks to processors by column, hence in our implementation, tasks that eliminate an entry in the first column, using the first row as a pivot, execute before any others. Again, the pivot row is used immediately and does not need to be stored any longer. This is not true of the medium-grain decomposition as implemented by US. Since a row defines the process granularity, a row can be allocated to a processor only when some previous row is completed. This sequence of operations requires that we

COMMUNICATION OPERATIONS REQUIRED BY VARIOUS DECOMPOSITION AND CACHING STRATEGIES

Processors	Small Grain	Small Grain (cache)	Medium Grain	Medium Grain (cache)	Large Grain (cache)
2	958800	640798	320400	2398	1598
4	958800	642396	320400	3996	3196
8	958800	645592	320400	7192	6392
16	958800	651984	320400	13584	12784
32	958800	664768	320400	26368	25568
64	958800	690336	320400	51936	51136
128	958800	741472	320400	103072	102272
256	958800	843744	320400	205344	204544

TABLE 1

ultimately store all pivot rows in the cache for use by the last row to execute. In effect, the caching strategy we have described is rendered impractical by our decomposition strategy in this case.

5. Implementation Performance

All of the decomposition and caching strategies were implemented, except the medium-grain decomposition with caching. (As was stated previously, the cache size required for the medium-grain decomposition is prohibitive.) Figure 1 summarizes the performance of the various strategies. The performance of the implementations is consistent with the data in Table 1. That is, the large-grain decomposition generates the least amount of communication and is the most efficient. The caching strategy improves the small-grain decomposition by a nontrivial amount, although the shape of both curves is the same. The medium-grain decomposition is a further improvement on the small-grain decomposition.

It is important to note that the number of communication operations required by a decomposition is not the sole factor in performance. The cost of a single communication operation varies depending on whether shared memory or message passing is used. The shared-memory operations are subsumed in the message-passing operations, which also include buffering and synchronization. Each message-passing operation is up to 5 times more expensive. For this reason, the disparity in performance is not as great as that suggested by Table 1.

For the problem size chosen, an 800x800 matrix, none of the implementations were able to exploit significantly more than 64 processors. Furthermore, the performance of the message-passing implementation actually deteriorates as additional processors are

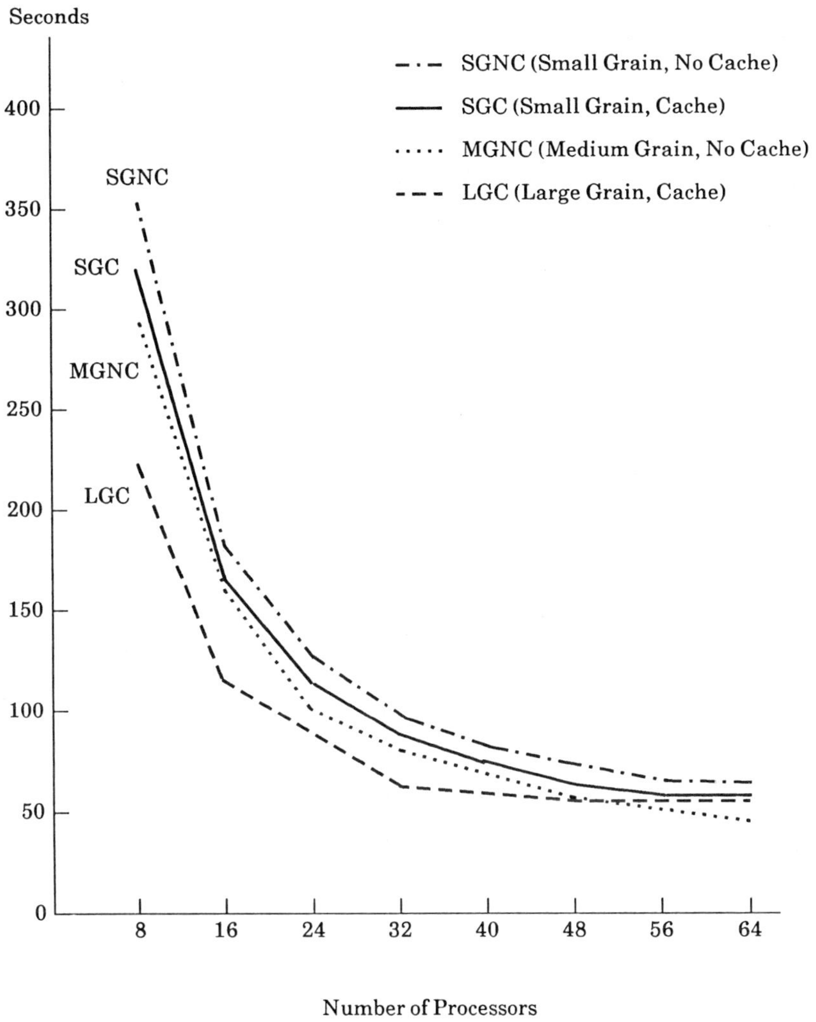

Implementation Performance on an 800 x 800 Matrix

Figure 1

used. Figure 2 contrasts the performance of the two extremes in implementation, small-grain tasks using shared-memory communication versus large grain processes communicating via messages. With four processors, the message-passing implementation is about 30% faster than the shared-memory implementation. The relative performance is consistent across different problem sizes and, in each case, the message-passing implementation is more efficient. The disparity decreases as additional processors are used until the "knee" in the performance curve of the message-passing system is reached.

The improved performance of the message-passing implementation can be directly attributed to data locality (*i.e.*, large-grain decomposition), which reduces the need for communication. Nearly every data access in the shared-memory implementation requires a remote reference. Even when the data physically resides in local memory, all data conceptually resides in shared memory and must be copied into the local workspace. Very few remote references are needed in the message-passing implementation (none, if only one processor is in use), and a copy takes place only when a remote reference is made. Nonetheless, communication via message passing is more expensive than communication via shared memory and each additional processor causes additional communication. For a fixed N, there exists a point of diminishing returns as P is increased. Once we reach the "knee" in the curve (approximately 64 processors), the additional communication required as processors are added to the work force is not offset by a significant gain in parallelism

To confirm the hypothesis that communication costs are the reason for the "knee" in the curve, the cost of sending a message was artificially varied to see what effect communication has on the overall performance of the message-passing implementation. The message-passing system was modified to perform 0, 1, 2, and 4 copy operations per message. As shown in Figure 3, the performance curve for message passing without a copy operation is similar to the curve for the shared-memory implementation. For a problem of size 200x200, a plateau is reached at 64 processors and additional processors neither help nor hinder. A single copy operation per message creates a "knee" at 26 processors. The "knee" moves down to 18 processors when two copy operations are made per message and falls even farther, to 13 processors, when four copy operations are used. This is precisely the effect we see in Figure 1, wherein the message-passing implementation reaches a "knee" at 64 processors, and then rapidly deteriorates in performance.

The shared-memory implementation does not exhibit a "knee" in our experiments because each additional processor does not significantly add to the communication costs of the implementation. In the small-grain decomposition without caching, the amount of communication is fixed, so additional processors do not add to the burden. In the small-grain decomposition with caching, each processor adds only slightly to the amount of communication, and unlike in the message-passing implementation, the cost of each additional communication operation is small. Any "knee" that might result would only appear well beyond the point for which we can gather empirical data.

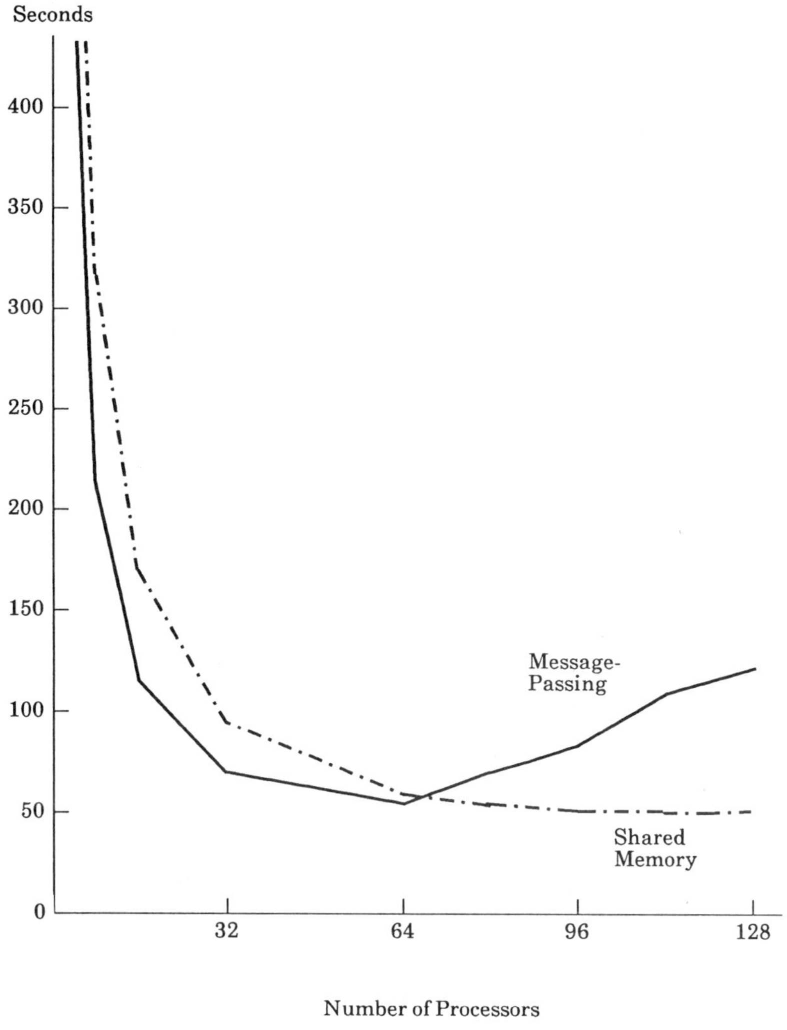

Shared Memory vs. Message-Passing
Implementation Performance on an 800 x 800 Matrix
Figure 2

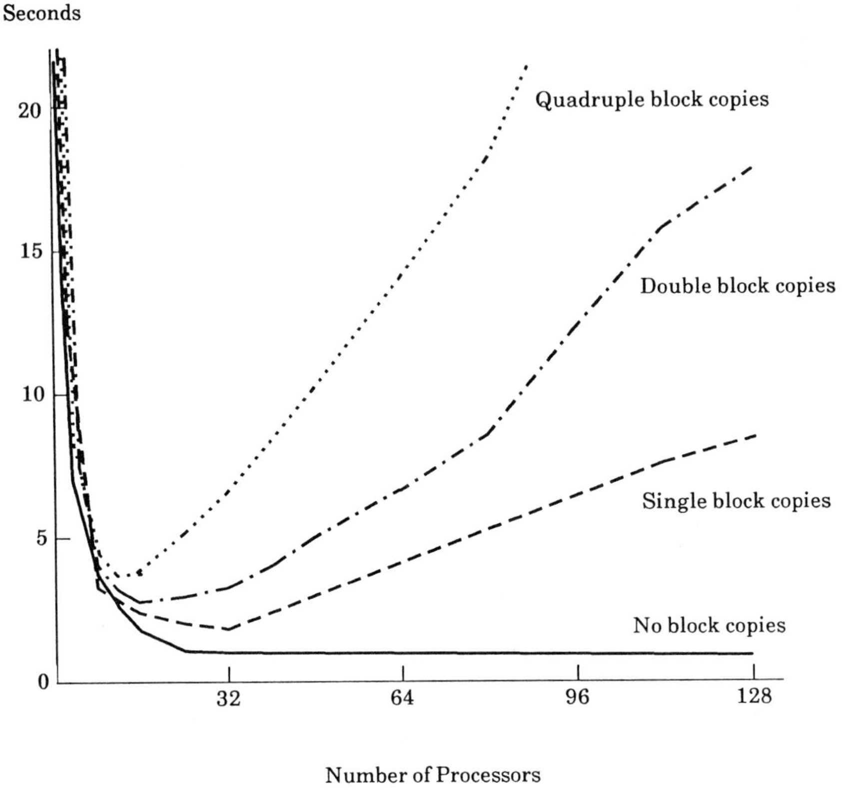

Message-Passing Communication Costs on a 200 x 200 Matrix

Figure 3

As a result, if P is significantly smaller than N, as it should be to achieve high processor efficiency, the message-passing implementation will make many fewer copies than the shared-memory implementation, although each copy (message) will take longer. Fewer copies will reduce both switch and memory contention, and improve performance, until the "knee" in the curve is reached.

5.1. Memory Contention

In the Uniform System implementations, all memories in the system are used to store the problem matrix, even when only a few processors are involved in the computation. This has the effect of improving the overall performance by decreasing memory contention. Since the message-passing implementation can not take advantage of memories in unused processors, it is important to quantify this effect when comparing implementations.

In order to determine the effect extra memories have on the performance of a computation, the small-grain decomposition with caching was executed on various numbers of processors and memories. The test results show that the extra memories can have a substantial impact on performance. On a problem matrix of size 400x400, a Butterfly configuration with 4 columns of switches, 16 processors, and 96 memories performed 15% better than 16 processors and 16 memories. On a problem matrix of size 200x200 and a Butterfly configuration with 2 columns of switches, 4 processors and 16 memories performed 30% better than 4 processors and 4 memories. The greatest effect occurs when roughly 1/4 to 1/2 of the total number of processors are in use. When a larger fraction of processors are performing computation, most of the memory is already in use, *i.e.*, there is no other extra memory. When too few processors are used, they are insufficient to generate enough load on the memory to make memory contention significant. Surprisingly, there was even a very small, but consistent, improvement of 1-2% in the performance of the single processor case, probably due to the fact that a block transfer from remote to local memory is slightly faster than a transfer within local memory. As extra memories were used, less data was retrieved from the local memory, slightly reducing the execution time.

5.2. Switch Contention

The 128-node Butterfly requires a switch configuration capable of supporting 256 nodes. The excess switch capacity can be used to provide redundant paths through the switch, increasing the total amount of potential bandwidth in the switch. Under software control, the Butterfly can also be configured to artificially limit the switch configuration to the bandwidth of a 64-port switch. We can use the existence of these alternative switch configurations to measure indirectly the switch contention in a program. That is, if a program does not exhibit significant switch contention, the performance of the program will not be seriously affected by the use of the smaller switch configuration.

The small-grain decomposition with caching was executed on a 128-node Butterfly using both switch configurations. The results of this experiment show that the effect of reduced switch capacity can be dramatic (as a percentage of total execution time) when a large number of processors are in use. A 35% improvement in execution speed was attained using the larger switch configuration for 94 processors on a problem of size 400x400; 64 processors achieved a speedup of more than 20% on the same problem using the larger configuration. Although these figures are not an exact measure of switch contention, they do suggest that the small-grain decomposition is generating traffic sufficient to cause significant switch contention. It is not surprising that this implementation introduces significant switch contention once the inner loop has been tuned to the point where the program is communication intensive.

The large-grain decomposition exploits the locality of data, thereby minimizing switch contention. Unlike the small-grain decomposition, the message-passing implementation for Gaussian elimination showed no significant switch contention. A problem of size 800x800 was able to execute on 32 processors at roughly the same speed using either switch configuration. Over the entire range of experiments, the effect of alternate communication paths improved the execution time of the message-passing implementation by at most 3%.

6. Conclusions

The intention of our experiments was to analyze the effect that the architecture and programming environment of a shared-memory multiprocessor have on problem decomposition and communication tradeoffs. Our experiments were designed to provide some empirical data that suggests how best to structure communication between processes in such an environment. Based on our experiences, we offer the following conclusions.

The particular model of computation in use is less important than how well it is matched to the application. Gaussian elimination is not a typical application for loosely-coupled systems; a multiprocessor is more appropriate. However, this does not mean that the shared-memory model is the best approach for this application. Since Gaussian elimination is essentially *value-oriented* (no addresses are communicated and rows are not used until they have stabilized), message passing is a better model of the communication that takes place. Obviously, there are many other applications for which message passing would be inappropriate. Any programming environment that offers a single model of communication will not be well-matched to a large class of applications.

A high-level interface, efficiently implemented, leaves the programmer fewer opportunities to introduce inefficiency. Of course this assumes that the high-level interface provides a *useful* abstraction. Communication via shared memory does not include synchronization primitives, although US does provide a busy-wait LOCK and UNLOCK primitive. Each programmer must use these to implement the appropriate synchronization. Spin locks are commonly used, but the resultant program can be especially sensitive to the amount of time spent between attempts to set the lock [8]. The only process synchronization in the message-passing implementation occurs during access to message buffers, which is implemented very efficiently using microcoded atomic operations. The user does not need to be concerned with low-level synchronization; it is implicit in the message-passing primitives.

Finally, *the performance of an application depends not only on the efficiency of the underlying communication, but also on the extent to which the underlying model of computation encourages or discourages communication.* The Uniform System model does not encourage the programmer to exploit locality. Data is assumed to reside in one large, globally-shared address space and the boundaries between processors are essentially ignored. Thus, in Gaussian elimination, each task must copy rows back and forth between local memory and globally-shared memory, independent of the location of the task and data. This is not the case with the message-passing model, since all data is local to some process in the computation. Locality makes it possible to avoid copying any row to be modified (since only local rows are ever modified) and also to avoid copying any pivot row that happens to be local. Thus, an implementation based on very efficient communication (*e.g.*, shared memory) may perform worse than one based on a less efficient mechanism (*e.g.*, message passing), if such efficiency requires or even encourages too much communication.

Acknowledgements

I am grateful for the efforts of the people who helped perform these experiments. Lawrence Crowl and Neal Gafter implemented various versions of the message-passing system and provided valuable insights. Thomas Olson performed many of the Uniform System experiments. Brian Marsh made several modifications to the Uniform System implementation and helped with the analysis. Liud Bukys provided technical assistance. I would also like to thank the members of the technical staff at BBN Laboratories for their comments and assistance, especially Bob Thomas, who was kind enough to share his data and program code. This work was supported by the National Science Foundation under grant number DCR-8320136 and by DARPA/ETL under research contract DACA76-85-C-0001.

References

1. Alliant Computer Systems Corporation, ALLIANT FX/Series, Product Summary (Unpublished), June 1985.
2. BBN Laboratories, The Uniform System Approach To Programming the Butterfly Parallel Processor, Version 1, Oct 1985.

3. W. Crowther, J. Goodhue, E. Starr, R. Thomas, W. Milliken and T. Blackadar, Performance Measurements on a 128-Node Butterfly Parallel Processor, *Proceedings of 1985 International Conference on Parallel Processing*, August 1985, 531-540.
4. T. J. LeBlanc, N. M. Gafter and T. Ohkami, SMP: A Message-Based Programming Environment for the BBN Butterfly, Butterfly Project Report 8, Computer Science Department, University of Rochester, July 1986.
5. W. Milliken et al, Chrysalis Programmer's Manual, Version 2.2, BBN Laboratories, June 1985.
6. V. Pan and J. Reif, Efficient Parallel Solution of Linear Systems, *Proc. 17th ACM Symp. on Theory of Computing*, Providence, Rhode Island, May 6-8, 1985.
7. Sequent Computer Systems Inc., Balance 8000 System, Technical Summary, Dec 1985.
8. R. Thomas, Using the Butterfly to Solve Simultaneous Linear Equations, *Butterfly Working Group Note 4*, BBN Laboratories, Mar 1985.

Domain Decomposition Preconditioners for Elliptic Problems in Two and Three Dimensions

JOSEPH E. PASCIAK

Brookhaven National Laboratory
Upton, N.Y. 11973

Abstract. In this talk, we shall describe some techniques for developing domain decomposition preconditioners for elliptic boundary value problems in two and three dimensions. We consider the case where more than two subdomains meet at an interior point of the original domain; this allows a subdivision into an arbitrary number of subdomains without the deterioration of the iterative convergence rates for the resulting algorithm. We set up a general framework for the development of preconditioners by domain decomposition. As examples of the application of these techniques, we derive domain decomposition preconditioners for elliptic problems in two and three dimensions.

1. INTRODUCTION

The need for modeling more complex physical processes has led to the development of larger and faster computers. In the next generation of machines, parallel computing architectures will be employed to gain additional computational improvement. If significant computational improvements are to be realized, then algorithms especially tailored to parallel environments must be developed. Moreover, these algorithms should be effective on machines with a large number of processors.

Preconditioners based on domain decomposition gives rise to an important approach for the development of parallel algorithms for elliptic boundary value problems. Such algorithms are straightforward to efficiently implement on actual parallel computers. In these implementations, a greater number of subdomains gives rise to a greater number of independent parallel tasks leading to an effective use of parallel resources. The overall efficiency of the resulting algorithm also depends upon the rate of iterative convergence which can be estimated in terms of the condition number of the preconditioned system. Accordingly, to be effective in a parallel environment, the conditioning of the preconditioned system should not deteriorate as the number of subregions (i.e. processors) increase. Thus, in this talk, I will only consider domain decomposition methods whose conditioning improves with the number of subdomains. Other domain decomposition preconditioners have been proposed [**1**,2,3,4,5,6,7,8].

In order to keep the notation as simple as possible, we will restrict our attention to the following model problem:

$$-\Delta u = f \text{ in } \Omega, \qquad (1.1)$$
$$u = 0 \text{ on } \partial\Omega.$$

This manuscript has been authored under contract number DE-AC02-76CH00016 with the U.S. Department of Energy. Accordingly, the U.S. Government retains a non-exclusive, royalty-free license to publish or reproduce the published form of this contribution, or allow others to do so, for U.S. Government purposes. This work was also supported in part under the National Science Foundation Grant No. DMS84-05352 and under the Air Force Office of Scientific Research, Contract No. ISSA86-0026.

Here Δ denotes the Laplace operator and Ω is a polygonal or polyhedral domain in R^N for $N = 2$ or 3. Extensions of the methods to more general problems on more general domains are possible [**5**,**6**].

Let $H_0^1(\Omega)$ denote the set of functions which vanish on $\partial\Omega$ and are in the Sobolev space $H^1(\Omega)$ (i.e. the functions which together with the first derivatives are square integrable on Ω). The weak form of (1.1) is then: Find $U \in H_0^1(\Omega)$ such that

$$D(u,\theta) = (f,\theta) \qquad \text{for all } \theta \in H_0^1(\Omega), \tag{1.2}$$

where

$$D(\varsigma,\theta) \equiv \int_\Omega \nabla\varsigma \cdot \nabla\theta \, dx$$

and

$$(\varsigma,\theta) \equiv \int_\Omega \varsigma\theta \, dx.$$

In the usual finite element approach, we are given a finite dimensional subspace S_h of $H_0^1(\Omega)$. The parameter h is related to the accuracy of the approximation and is roughly the mesh size. The Galerkin approximation to u is then the unique function $U \in S_h$ satisfying

$$D(U,\omega) = (f,\omega) \qquad \text{for all } \omega \in S_h. \tag{1.3}$$

We shall be concerned in this talk with the efficient solution of (1.3) using preconditioners based on domain decomposition. The question of defining a preconditioner for (1.3) may be approached from two points of view. The first requires the choice of a basis for S_h. Employing this basis, one is led to a matrix problem for the computation of the corresponding coefficients of U. The preconditioning problem from this point of view is to define another matrix which is easier to invert and 'spectrally close' to the original. Alternatively, the preconditioning problem can be written as a problem of defining a symmetric positive definite quadratic form $B(\cdot,\cdot)$ which is equivalent to $D(\cdot,\cdot)$ on $S_h \times S_h$. At each step of the iteration we must solve problems of the form: Given a linear functional G on S_h, find $W \in S_h$ such that

$$B(W,\omega) = G(\omega) \qquad \text{for all } \omega \in S_h. \tag{1.4}$$

The problem of finding the solution of (1.4) should be computationally less complex than that of finding the solution to (1.3) on the given computer architecture. The corresponding spectral condition in terms of forms reduces to inequalities of the form

$$c_0 B(\omega,\omega) \le D(\omega,\omega) \le c_1 B(\omega,\omega) \qquad \text{for all } \omega \in S_h. \tag{1.5}$$

The condition number of the preconditioned system is bounded by c_1/c_0. In the above inequality, c_0 and c_1 are constants which may depend on d (the subdomain size) and h.

In the remainder of the talk, we shall use generic constants of the form C_i to denote constants which do not depend on d and h. These constants may take on different values in different places.

2. A GENERAL DOMAIN DECOMPOSITION APPROACH

In this section, we shall set up a general technique for developing domain decomposition preconditioners. In later sections, we shall us this technique to derive domain decomposition preconditioners for elliptic problems in R^2 and R^3.

Although the algorithms and analysis extend to more complicated elements, in the talk we shall only consider the simplest approximation subspaces. For the two dimensional problem, we consider continuous piecewise linear functions on a 'quasi-uniform' triangulation of size h. In R^3, we subdivide Ω into rectangular prisms $\Omega = \cup \tau_j$ and define S_h to be the set of functions which are continuous on Ω and piecewise tri-linear on the τ_j.

To define domain decomposition algorithms, we naturally must start with a decomposition of Ω into M subdomains, $\Omega = \cup_{i=1}^{M} \Omega_i$. The hypothesis on these subdomains and their relation to the discrete spaces S_h will not be given in full generality. Instead, for simplicity of presentation, subdomains in R^2 (resp. R^3) should be thought of as triangles or rectangles (resp. rectangular prisms). The subregions are to be 'quasi-uniform' of size d, i.e. they should be contained in a ball of radius $C_0 d$ and contain a ball of radius $C_1 d$. Furthermore, the boundary of the subdomains should be part of the mesh boundary of the subspace S_h. This means that any given mesh triangle or rectangular prism in the subspace definition is contained in some $\bar{\Omega}_j$.

To define the domain decomposition algorithms, we proceed as follows: Let $\Gamma = \cup \partial \Omega_j$ and define

$$S_h^0 = \{\omega \in S_h | \omega = 0 \text{ on } \Gamma\}. \tag{2.1}$$

We decompose arbitrary functions $W \in S_h$ by $W = W_P + W_H$ where $W_P \in S_h^0$ and

$$D(W_H, \omega) = 0 \qquad \text{for all } \omega \in S_h^0. \tag{2.2}$$

W_H is the unique function in S_h which equals W on Γ and satisfies (2.2). Such a function will be called 'discrete harmonic.' It is an immediate consequence of (2.2) that

$$D(W, W) = D(W_P, W_P) + D(W_H, W_H). \tag{2.3}$$

A central theme in the derivation of domain decomposition preconditioners is to define $B(\cdot, \cdot)$ by replacing the form $D(W_H, W_H)$ in (2.3). Notice that W_H implicitly only depends on its values on the nodes on Γ. Thus, it is reasonable to replace $D(W_H, W_H)$ with a form which explicitly only depends upon the boundary values of W_H. A central issue in this subject is the appropriate selection of the replacement form.

We shall derive replacement forms in the following way. We obviously have that

$$D(W_H, W_H) = \sum_{i=1}^{M} D_i(W_H, W_H), \tag{2.4}$$

where

$$D_i(W_H, W_H) \equiv \int_{\Omega_i} |\nabla W_H|^2 \, dx.$$

We shall define B by replacing the terms of (2.4) on a subdomain by subdomain basis.

To do this, we first consider the weighted one-norm on the subdomains given by

$$\|v\|_{1,\Omega_i}^2 \equiv D_i(v,v) + d^{-2}\|v\|_{L^2(\Omega_i)}^2 . \tag{2.5}$$

We assume that we are given forms $A_i(\cdot,\cdot)$ which are defined on $S_h|_{\partial\Omega_i} \times S_h|_{\partial\Omega_i}$ satisfying

$$\alpha_0(d/h)\, A_i(V,V) \le \|V\|_{1,\Omega_i}^2 \le \alpha_1(d/h)\, A_i(V,V), \tag{2.6}$$

for all $V \in S_h|_{\Omega_i}$ with V discrete harmonic on Ω_i. Here $\alpha_0(d/h)$ and $\alpha_1(d/h)$ are functions of d/h. The form of the weighted norm in (2.6) is quite appropriate. In fact, if one maps the subregion to a unit size domain $\tilde{\Omega}$ via a linear mapping $T(\Omega) = \tilde{\Omega}$, then (2.6) is equivalent to

$$\alpha_0(d/h)\, A_i(V,V) \le d^{N-2} \left\|\tilde{V}\right\|_{H^1(\tilde{\Omega})}^2 \le \alpha_1(d/h)\, A_i(V,V),$$

where $\tilde{V} = V \circ T^{-1}$. If we set $\tilde{A}_i(\tilde{V},\tilde{V}) \equiv A_i(V,V)$, then the question of defining A_i reduces to that of defining an appropriate $\tilde{A}_i$ on the unit size domain. Note that d/h is the mesh size for the mapped subspace. Thus, the problem of constructing and analyzing the forms $A_i(\cdot,\cdot)$ reduces to corresponding problems on unit size domains (independent of the subdomain parameter d). We shall give examples of these constructions in later sections.

However, it does not simply suffice to replace the terms $D_i(W_H,W_H)$ with $A_i(W_H,W_H)$. Clearly, $D_i(\cdot,\cdot)$ is indefinite since it annihilates constants. Consequently, it should be replaced with a form which also annihilates constants. Accordingly, we modify A_i so that the new form is zero for constants and thus define

$$B(W,W) = D(W_P,W_P) + \sum_{i=1}^{M} A_i(W_H - \bar{W}_H^i, W_H - \bar{W}_H^i), \tag{2.7}$$

where $\bar{W}_H^i$ is the constant on Ω_i which satisfies

$$A_i(W_H - \bar{W}_H^i, 1) = 0. \tag{2.8}$$

One can easily prove the following theorem.

THEOREM 1. *There are constants β_0 and β_1 which do not depend on d or h satisfying*

$$\beta_0\alpha_0(d/h)\, B(W,W) \le D(W,W) \le \beta_1\alpha_1(d/h)\, B(W,W), \qquad \textit{for all } W \in S_h. \tag{2.9}$$

REMARK 1: It is important, but not necessary, to use (2.7)-(2.8) to properly treat the constants. One could use instead,

$$B(W,W) = D(W_P,W_P) + \sum_{i=1}^{M} A_i(W_H,W_H). \tag{2.10}$$

The solution of (1.4), when B is given by (2.10), is somewhat easier to compute than when B is given by (2.7). However, in all of the applications to be discussed later, the condition number of the preconditioned system using (2.10) is d^{-2} times larger than the corresponding condition number using (2.7).

3. A TWO DIMENSIONAL APPLICATION

In this section, we develop a two dimensional domain decomposition preconditioner using the techniques of Section 2. Using the analysis given in [**5**], it can be shown that the condition number for the resulting preconditioned system is bounded by $C(1+\ln^2(d/h))$. Hence, the method presented here should be competitive with that given in [**5**].

From the discussion in Section 2, it clearly suffices to define the forms $\tilde{A}_i(\cdot,\cdot)$ on the reference size domain $\tilde{\Omega}$. Let us assume that the subdomains are rectangular; the triangular case is similar. Thus the reference domain is rectangular. Let $\tilde{S}_h$ denote the image of S_h under T, i.e.

$$\tilde{S}_h \equiv \{V \circ T^{-1} | V \in S_h\}.$$

The space $\tilde{S}_h$ clearly consists of a set of functions which are piecewise linear on the image of the triangulation defining S_h. The triangulation defining $\tilde{S}_h$ is quasi-uniform of size d/h. It is also clear that the image $\tilde{V} \in \tilde{S}_h$ of a discrete harmonic function $V \in S_h$ is discrete harmonic with respect to $\tilde{S}_h$. For such functions, it is well known (see, for example, [**4**,5]) that

$$C_0 \left|\tilde{V}\right|^2_{H^{1/2}(\partial\tilde{\Omega})} \le \left\|\tilde{V}\right\|^2_{H^1(\tilde{\Omega})} \le C_1 \left|\tilde{V}\right|^2_{H^{1/2}(\partial\tilde{\Omega})}.$$

The problem then is define computationally feasible replacements for $|\cdot|^2_{H^{1/2}(\partial\tilde{\Omega})}$.

To define the replacement, we will use the discrete operator $l_0^{1/2}$. Let Γ_i be an edge of $\tilde{\Omega}$ and

$$\tilde{S}_h(\Gamma_i) \equiv \{\phi|_{\Gamma_i} \text{ such that } \phi \in \tilde{S}_h \text{ and } \phi = 0 \text{ at the corners of } \tilde{\Omega}\}.$$

The discrete operator $l_0 : \tilde{S}_h(\Gamma_i) \mapsto \tilde{S}_h(\Gamma_i)$ is defined by $l_0\theta = \eta$ where η solves

$$\int_{\Gamma_i} \eta\omega \, dx = \int_{\Gamma_i} \theta'\omega' \, dx \qquad \text{for all } \omega \in \tilde{S}_h(\Gamma_i). \tag{3.1}$$

The operator l_0 is symmetric positive definite on $\tilde{S}_h(\Gamma_i)$ and $l_0^{1/2}$ is defined to be its square root.

We then define the form $\tilde{A}_i$ by

$$\tilde{A}_i(\tilde{V},\tilde{V}) \equiv \sum_{x_j \text{ corners}} \tilde{V}(x_j)^2 + \sum_{\Gamma_i} \left\langle l_0^{1/2}\tilde{V},\tilde{V}\right\rangle_\Gamma . \tag{3.2}$$

Here $\langle\cdot,\cdot\rangle_{\Gamma_i}$ denotes the L^2 inner product on Γ_i and the second sum is over all edges of $\tilde{\Omega}$. Using Theorem 1 and the techniques of [**5**], we can prove the following theorem.

THEOREM 2. *Let B be given by (2.7) with A_i defined by*

$$A_i(V,V) = \tilde{A}_i(\tilde{V},\tilde{V}) \tag{3.3}$$

where $\tilde{A}_i(\cdot,\cdot)$ is given by (3.2). Then there are constants C_0 and C_1 which are independent of d and h and satisfy

$$C_0 D(W,W) \le B(W,W) \le C_1(1+\ln^2(d/h))D(W,W) \qquad \text{for all } W \in S_h.$$

REMARK 2: The preconditioner defined by (2.10) and (3.2) was proposed by M. Dryja. Using the techniques given in [5], it is possible to show that the condition number for the resulting method is bounded by $Cd^{-2}(1+\ln^2(d/h))$. This method may not perform well in parallel applications where a large number of subdomains must be used since the computational improvement attained from parallelization may be offset by the increased number of iterations required for convergence.

4. THE THREE DIMENSIONAL EXAMPLES

In this section, we develop two three dimensional domain decomposition preconditioners using the techniques of Section 2. The first example was given in more generality in [6] and leads to a preconditioned system with condition number bounded by Cd/h. The second example is new and gives rise to a preconditioned system with condition number bounded by $C(1+\ln^2(d/h))$.

There is another three dimensional domain decomposition method developed by Bramble, Pasciak and Schatz which also gives rise to a condition number growth of $C(1+\ln^2(d/h))$. That method is similar to the two dimensional scheme described in [5]. We shall not discuss it here since it does not fit into the framework of Section 2.

The first method that we shall consider is the one described in [6]. We define

$$\tilde{A}_i(\tilde{V},\tilde{V}) \equiv h \sum_{x_j\in\partial\tilde{\Omega}} \tilde{V}(x_j)^2. \tag{4.1}$$

The sum in (4.1) is taken over all nodes x_j on $\partial\tilde{\Omega}$. The following theorem is proved in [6].

THEOREM 3. *Let B be given by (2.7) with A_i defined by (3.3) and $\tilde{A}_i(\cdot,\cdot)$ given by (4.1). Then there are constants C_0 and C_1, which are independent of d and h, satisfying*

$$C_0 D(W,W) \le B(W,W) \le \frac{C_1 d}{h} D(W,W) \qquad \text{for all } W\in S_h.$$

As in the two dimensional case, to define a 'log squared' method, we must introduce a discrete operator $l_0^{1/2}$. Let Γ_i denote a face of the rectangular domain $\tilde{\Omega}$ and define

$$\tilde{S}_h(\Gamma_i) \equiv \{\phi|_{\Gamma_i} \text{ with } \phi\in\tilde{S}_h \text{ and } \phi=0 \text{ on the remaining faces of } \tilde{\Omega}\}.$$

The discrete operator $l_0 : \tilde{S}_h(\Gamma_i) \mapsto \tilde{S}_h(\Gamma_i)$ is then defined by $l_0\theta = \eta$ where η is the solution to

$$\int_{\Gamma_i} \eta\omega\,dx = \int_{\Gamma_i} \nabla\theta\cdot\nabla\omega\,dx \qquad \text{for all } \omega\in\tilde{S}_h(\Gamma_i).$$

Once again, $l_0^{1/2}$ is defined to be the square root of l_0.

The boundary of $\tilde{\Omega}$ can be partitioned into interiors (in a two dimensional sense) of the faces and the edges and corners. We shall consider the set of points on the edges and corners to be the 'wire box.' Let $\tilde{S}_h(\Gamma)$ denote $\tilde{S}_h$ restricted to $\partial\tilde{\Omega}$. We decompose functions $\tilde{V} \in \tilde{S}_h(\Gamma)$ into $\tilde{V} = \tilde{V}_F + \tilde{V}_E$ where $\tilde{V}_F$ vanishes on the wire box and $\tilde{V}_E$ is zero on the boundary nodes which are interior to the faces. Using this decomposition, we define

$$\tilde{A}_i(\tilde{V},\tilde{V}) \equiv h \sum_{x_j \in \text{ wire box}} \tilde{V}_E(x_j)^2 + \sum_{\Gamma_i} \left\langle l_0^{1/2}\tilde{V}_F, \tilde{V}_F \right\rangle_{\Gamma_i}. \tag{4.2}$$

Here the first sum is over the nodes x_j in the wire box, the second sum is over the faces of $\partial\tilde{\Omega}$, and $\langle\cdot,\cdot\rangle_{\Gamma_i}$ denotes the L^2 inner product on Γ_i. The proof of the following theorem will be given in a subsequent paper.

THEOREM 4. *Let B be given by (2.7) with A_i defined by (3.3) and $\tilde{A}_i(\cdot,\cdot)$ given by (4.2). Then there are constants C_0 and C_1, which are independent of d and h satisfying*

$$C_0 D(W,W) \le B(W,W) \le C_1(1+\ln^2(d/h))D(W,W) \qquad \textit{for all } W \in S_h.$$

REMARK 3: In both of the above examples, it is important to subtract out the constants on the subdomains as described by (2.7) and (2.8). Indeed, if one used (2.10) with (4.1) then the condition number for the preconditioned system would grow like $\frac{C}{hd}$. Similarly, if one used (2.10) with (4.2), the condition number would grow like $Cd^{-2}(1+\ln^2(d/h))$. In both cases, the condition number and hence the number of iterations required for convergence would grow with the number of subdomains used.

5. COMPUTING THE ACTION OF THE INVERSE OF B

We shall briefly discuss the problem of computing the action of the inverse of the preconditioner in this section. In general, when B is of the form (2.7) (or (2.10)), to solve (1.4), we first find W_P, next compute the values of W_H on Γ, and finally extend W_H to all of Ω. This process can be also interpreted in terms of block Gaussian elimination.

We now give the details of this three step algorithm for the solution of (1.4). As already mentioned, the problem of finding the solution W to (1.4) reduces to that of computing W_P and W_H. The first step is to compute W_P. By taking $\Phi \in S_h^0$ in (2.7), we see that

$$B(W_P,\Phi) = G(\Phi) \qquad \text{for all } \Phi \in S_h^0. \tag{5.1}$$

Equation (5.1) shows that W_P can be determined by solving independent discrete Dirichlet problems on the subregions. The second step involves the computation of the values of W_H on Γ. These values are determined as the solution of the following problem:

$$A_i(W_H - \bar{W}_H^i, \theta) = G(\tilde{\theta}) - A_i(W_P,\tilde{\theta}) \qquad \text{for all } \theta \in S_h(\Gamma). \tag{5.2}$$

Here $S_h(\Gamma)$ is S_h restricted to Γ and $\tilde{\theta}$ denotes any extension of θ in S_h. The development of an algorithm for solving (5.2) is an essential part of the solution process. We shall give some indication how (5.2) can be solved; however, a complete description is beyond the scope of this talk. The third step is to compute the discrete harmonic extension of the boundary values of W_H computed in the previous step. This step also reduces to independent discrete Diriclet problems on the subdomains which can be solved in parallel (see [**5**,**6**] for details).

We next indicate how the boundary values of W_H solving (5.2) can be economically computed. We first observe that W_H on Γ can be easily computed if the values of $\bar{W}_H^i, i = 1, \ldots, M$ are known. Indeed, by moving the terms involving $\bar{W}_H^i$ to the right hand of (5.2), we are left with a problem of the form

$$A_i(W_H, \theta) = F(\theta) \qquad \text{for all } \theta \in S_h(\Gamma), \tag{5.3}$$

where F is an appropriate (known) linear functional. If $A_i(\cdot,\cdot)$ is given by (4.1) and (3.3), then the solution of (5.3) only involves the inversion of a weighted identity matrix. If $A_i(\cdot,\cdot)$ is given by (4.2) and (3.3), then the solution of (5.3) involves the inversion of a weighted identity matrix for the nodes on the wire basket and the inversion of the $l_0^{1/2}$ operator on the faces. A similar observation holds for the two dimensional example.

Thus, we could efficiently solve (5.2) if we could compute the values of $\bar{W}_H^i$. It is not at all obvious that this can be done in an efficient way. However, by an appropriate choice of basis, it is possible to derive a sparse, symmetric positive definite system of equations for these values. The description of this basis will not be given here. It is fully described in [6] for (4.1) and can be easily extended for (4.2) or (3.2). Thus, we propose solving (5.2) by first solving the sparse $M \times M$ system for the average values $\bar{W}_H^i$ and then solving (5.3) to compute W_H on Γ.

6. Numerical examples

In this section, we give some numerical examples which indicate that the domain decomposition preconditioners behave as theoretically predicted. We will only provide numerical results for the three dimensional applications of Section 4.

All of numerical experiments are applied to the standard model problem given by (1.1) where Ω is the unit cube. There are many techniques available for solving this problem. However, this problem is interesting in that it illustrates many of the convergence properties of the proposed preconditioner.

The first table gives results for the condition number K of the preconditioned system as a function h. For this example, the preconditioner is defined using (2.7) with A_i given by (3.3) and $\tilde{A}_i(\cdot,\cdot)$ given by (4.1). There were twenty seven subdomains of equal size so $d = 1/3$ remained fixed. We also report n, the number of iterations required to reduce the H^1 error by a factor of .0001 and ρ_0, the average reduction per iteration in that norm. As predicted by Theorem 3, the condition number for the preconditioned system grows like cd/h.

h	K	ρ_0	n
1/6	6.8	.39	11
1/12	17.4	.55	16
1/24	38	.64	21

Table 6.1. Iterative Convergence Results with Preconditioner Given by (4.1),(3.3) and (2.7).

For the next example, we use exactly the same problem and subdomains as the previous. However, we consider the preconditioner defined using (2.7) with A_i given by (3.3) and

$\tilde{A}_i(\cdot,\cdot)$ given by (4.2). Even though the results for $h = 1/6$ are slightly worst than those of Table 6.1, Table 6.2 illustrates the slower asymtotic growth associated with the 'log squared' estimate of Theorem 4. Note that Table 6.2 gives better results for smaller h.

h	K	ρ_0	n
1/6	11.8	.48	13
1/12	15	.51	14
1/24	19	.52	14

Table 6.2. Iterative Convergence Results with Preconditioner Given by (4.2),(3.3) and (2.7).

ACKNOWLEDGMENTS

The results reported in this talk represent joint work with James H. Bramble and Alfred H. Schatz of Cornell University, Ithaca, N.Y.

REFERENCES

1 P.E. Bjørstad and O.B. Widlund, *Solving elliptic problems on regions partitioned into substructures*, "Elliptic Problem Solvers II," (eds, G. Birkhoff and A. Schoenstadt), Academic Press, New York, 1984, pp. 245–256.

2 P.E. Bjørstad and O.B. Widlund, *Iterative methods for the solution of elliptic problems on regions partitioned into substructures*, (preprint).

3 J.H. Bramble, R.E. Ewing, J.E. Pasciak and A.H. Schatz, *A preconditioning technique for the efficient solution of problems with local grid refinement*, Comp. Meth. Appl. Mech. Eng. (to appear).

4 J.H. Bramble, J.E. Pasciak and A.H. Schatz, *An iterative method for Elliptic problems on regions partitioned into substructures*, Math. Comp. **46** (1986), 361–369.

5 J.H. Bramble, J.E. Pasciak and A.H. Schatz, *The construction of preconditioners for elliptic problems by substructureing, I*, Math. Comp. **47** (1986), 103–134.

6 J.H. Bramble, J.E. Pasciak and A.H. Schatz, *The construction of preconditioners for elliptic problems by substructuring, II*, Math. Comp. (to appear).

7 Q.V. Dihn, R. Glowinski and J. Périaux, *Solving elliptic problems by domain decomposition methods*, in "Elliptic Problem Solvers II," (eds, G. Birkhoff and A. Schoenstadt), Academic Press ', New York, 1984, pp. 395–426.

8 G.H. Golub and D. Meyers, *The use of preconditioning over irregular regions*, "Proceedings of the Sixth International Conference on Computing Methods in Science and Engineering" (to appear).

METHODS FOR AUTOMATED PROBLEM MAPPING

Joel Saltz*
Research Center for Scientific Computation
P.O. Box 2158 Yale Station
Yale University
New Haven CT 06520

Abstract.

It is anticipated that in order to make effective use of many future high performance architectures, programs will have to exhibit at least a medium grained parallelism. Methods for aggregating work represented by a directed acyclic graph are of particular interest for use in conjunction with techniques now under development for the automated exploitation of parallelism.

In this paper we present a framework for partitioning very sparse triangular systems of linear equations that is designed to produce favorable performance results in a wide variety of parallel architectures. Efficient methods for solving these systems are of interest because (1) they provide a useful model problem for use in exploring heuristics for the aggregation, mapping and scheduling of relatively fine grained computations whose data dependencies are specified by directed acyclic graphs and (2) because such efficient methods can find direct application in the development of parallel algorithms for scientific computation.

Simple expressions are presented that describe how to schedule computational work with varying degrees of granularity. We use the Encore Multimax as a hardware simulator to investigate the performance effects of using the partitioning techniques presented here in shared memory architectures with varying relative synchronization costs.

1 Introduction

A set of techniques is proposed for producing parameterized mappings of the solution of a very sparse triangular system of linear equations onto a range of parallel architectures. The solution of very sparse triangular system in which matrices typically have just a few non-zero elements per row is a particularly interesting problem to investigate for a variety of reasons.

*Department of Computer Science, Yale University, New Haven, CT 06520. saltz@yale, Work supported in part by the Office of Naval Research under Contract No. N00014-86-K-0310 ,NSF grant DCR 8106181 and NASA contract NAS 1-18107

In solutions of very sparse triangular systems, the number of floating point operations that can be performed at any one time is typically rather limited, hence in this problem synchronization and communication overheads play a particularly crucial role in determining the performance that can be achieved. The problem thus provides a simple yet challenging context in which to develop practical methods for automated problem mapping and work aggregating techniques. Efficient methods for solving very sparse triangular systems, easily adapted to a variety of architectures, have direct application in allowing allowing the efficient parallelization of Conjugate Gradient type algorithms preconditioned with incomplete LU factorizations. In our discussions and experimental investigations, we will focus in particular on sparse triangular systems that are generated from incomplete factorizations of matrices arising from discretizations of two dimensional partial differential equations. The work presented here is exerpted from the more detailed report [12].

Methods for aggregating work represented by a directed acyclic graph are of particular interest for use in conjunction with techniques under development for automating the exploitation of parallelism. In the context the ongoing Crystal/ACRE [14] parallel programming environment development effort, we are developing simple sets of techniques that can be used for the automated mapping of a variety of problems onto both very tightly coupled systems as well as onto systems that are quite loosely coupled. The study of methods for aggregating the work involved in solving very sparse triangular systems provides a tractable model system for the exploration of methods for aggregating work represented by various types of directed acyclic graphs arising during the compilation and execution of Crystal/ACRE programs.

In the Crystal/ACRE system, a very high level algorithm specification is supplied by the user. In this specification, the detailed interactions among processes in space and time are suppressed. The Crystal compiler and runtime system are being designed to allow the generation of instructions to direct an assemblage of communicating processes in the efficient execution of the specified algorithm. The compiler generates as many logical processes as possible and then the compiler or the runtime system combines clusters of logical processes to produce a problem decomposition that possesses a degree of granularity that is appropriate for the target machine.

In scientific applications, one frequently wishes to obtain multiple solutions with the same triangular system. Consequently it can be appropriate to spend a modest amount of time prescheduling the computations. Because of the small number of floating point computations required to solve each row, it is essential that very little time be spent during the algorithm's execution in scheduling row solutions. The approach taken here is to develop a set of methods requiring a set of standard preprocessing techniques that will allow this problem to be mapped or scheduled onto architectures with widely varying characteristics. The techniques will involve choosing parameters that allow one to make a variety of tradeoffs in the schedule or mapping specification.

We will assign to a single processor all computations pertaining to a row of the matrix. All computations pertaining to a given row are performed during a the first phase after the data required is known to be available. Note that this implies a potentially fine degree of granularity as we have a stated interest in matrices having few non zero elements in a row. The concurrency achievable through the use of this algorithm is determined by the dependencies between the rows of the triangular matrix. The computation is partitioned into phases and simple expressions are presented that describe the scheduling of computational work. This paper will for the most part discuss these methods in the context of architectures that support shared memory and consider the tradeoffs between synchronization delays and load balance. When appropriate, we will also discuss the use and performance of these mappings in environments that support fast preferential access to a local memory. The mapping techniques discussed here are currently being implemented on a message passing machine, the experimental results will be presented elsewhere.

In the execution of a fine grained problem on a shared memory machine such as the Encore Multimax, primary impediments to the achievement of ideal multiprocessor performance are

(1) load imbalance, (2) synchronization delays and (3) programming techniques that introduce computations not found in a corresponding sequential program intended to coordinate the parallel execution of a problem. Our techniques provide techniques that allow one to reduce (1) and (2). Since a prescheduled approach is used, it will be shown that it is possible to keep (3) from becoming a serious problem. The strategies developed here for dealing with these overheads are to: (A) reduce the number of synchronizations required during the solution of the problem, (B) make synchronizations less expensive and (C) improve the balance of load in between synchronizations. Due to the rather slow computation and the rapid synchronization, the Multimax presents a rather benign parallel environment. We will consequently also make use of this machine to provide hardware simulations of algorithm performance in architectures with relatively larger synchronization times.

In message passing environments (such as the Intel iPSC), (1) and (3) remain crucial impediments to the achievement of ideal multiprocessor performance. In current message passing machines communication startups are quite expensive ([13]); techniques that minimize the number of such startups are consequently of crucial importance. The techniques discussed here that reduce the number of synchronizations clearly also the number of startups required. In message passing machines the amount of information communicated also can play an important role in the determination of performance. As we shall see from the analysis of a model problem below, the specification of the parameters in the parameterized mapping also plays an important role in determining the amount of information that must be communicated.

The problem partitions and work schedules that result from the process described above may be viewed as a generalization of the work described by Saad [15]. In that report, a wavefront method was proposed for scheduling work involved in forward and backsolves of matrices arising from incomplete factorizations of matrices generated by 5 point discretizations of two dimensional elliptic partial differential equations. The work described by Saad as well as the results presented here assume a row oriented matrix storage scheme. Experimental work has been reported on the NYU ultra computer prototype involving the use of a wavefront method, where the work involved in solving for rows of sparse triangular linear systems was allocated in a self-scheduled manner [6].

A related body of literature also exists on the solution of triangular systems that are less sparse than the ones described here, these systems are generally obtained from matrix factorizations used in direct methods for solving sparse or non-sparse systems of linear equations. In these problems, the data dependencies between rows of the triangular matrix preclude the efficient use of methods that schedule all work pertaining to a given row at a time when all required data is available.

George [4] presents algorithms for a column oriented sparse cholesky factorization, along with algorithms for column oriented forward and backsolves. These algorithms utilize the notion of a pool of tasks whose parallel execution is controlled by a self-scheduling discipline. Heath [7] presents algorithms for parallel solution of triangular systems in distributed memory multiprocessors; these algorithms utilize a type of adjustable parameter for controlling algorithm granularity quite different from the ones discussed here. In the algorithms described in [7], the work required to calculate the inner products involved in solving for each row is shared among the processors. In very sparse triangular systems considered in this paper, there are very few computations involved in solving for a given variable. In these systems, parallelism can be obtained because the data dependencies between rows can allow one to solve for many variables simultaneously.

In section 2 we discuss methods for generating a parameterized problem decomposition that allows for considerable flexibility in determining the granularity of parallelism and facilitates inexpensive forms of synchronization. This defines a kind of coordinate system that can be used to advantage in specifying how the problem is to be solved. We present in section 3, expressions using the parameters arising from the decomposition defined in section 2 that allow the specification of the decomposition of the triangular system in a way that guarantees that the data dependencies in the problem will be respected. The expressions describing the parametrized schedules depend on, among other things, the type of synchronization utilized.

In section 4, through the analysis of a model problem, an analysis is made of the the tradeoffs between load imbalance and synchronization costs, as well as the tradeoffs between load imbalance and communication costs. An inexpensive method for explicitly balancing the load during each a phase of computation is described in section 5. In section 6 the results are reported of experimental investigations on the Encore Multimax multiprocessor that (1) explore the effect of parametric variations of granularity on performance, (2) evaluate the value of explicitly balancing load during each computational phase and (3) compare the performance obtained through the use of different synchronization techniques.

2 Problem Partitioning

2.1 Overview

The methods of problem partitioning described here have a strong geometrical motivation and will consequently first be introduced in a geometrical context. In the simplest form of incomplete LU preconditioning, the factors L and U have the same sparsity structure as the lower and upper portions of A respectively. A prior knowledge of the sparsity structure will be used to advantage in the generation of the following parameterized problem mapping. Note that this prior knowledge is not needed when the automated version of the problem mapping is used. This automated version of problem mapping will be described in the following section.

We will assume that we have a rectangular array of grid points, all points are connected with the same stencil. The stencil is assumed to link a given point with it's left, right, upper and lower neighbors in the grid. The matrix is formed by using the so called natural ordering in which grid points are numbered in a row-wise fashion beginning with the first column of the first row of the domain. We assume the same stencil is utilized for all mesh points in the problem.

The data dependency pattern between unknowns in the lower triangular solution may be best understood by referring back to the stencil and the grid utilized in the formulation of the problem [15]. Let $x_{i,j}$ be the location of a mesh point in the two dimensional domain, where $1 \leq i \leq n$ and $1 \leq j \leq n$. In the definition of the problem, a function value at a point $x_{i,j}$ is linearly dependent on function values at a given set of surrounding points. When a system involving a lower triangular matrix with the same sparsity structure as A is solved, the only interactions that need be considered are with variables in the grid that are in rows before i, as well as variables in row i that are before column j.

The grid points in a given row must be solved for sequentially, due to the coupling of each point to it's immediate neighbors. We assume that the stencil is rather small, so that relatively few calculations are involved in obtaining the value for a single grid point of the domain. In these mappings, the smallest unit of work that may be assigned to a particular processor consists of the computations pertaining to a particular row of grid points. The computations in a given row i depend only on results from row $j < i$. Depending on the relative size and properties of the problem and of the machine, better performance may be obtained by using a coarser grained assignment of work in which contiguous blocks of several rows are assigned to each of k processors. When there are more blocks of rows than there are processors, a wrapped assignment is used in which blocks are assigned to processors modulo k.

Given a fixed assignment of grid points to processors, one may be free to schedule the work associated with calculating values at mesh points in a variety of different ways. This processor scheduling has a marked effect on the frequency with which processors must interact to exchange

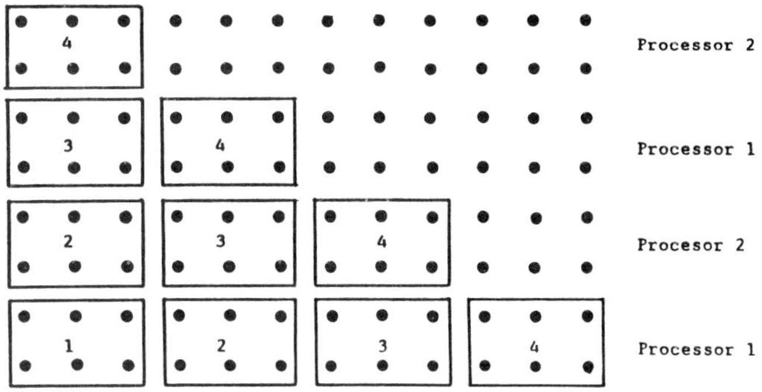

Figure 1: Two processors, five point stencil, block size = 2, window = 3. Numbers designate computational phases.

information. When a five point stencil is utilized, a convenient method of scheduling is to partition each block into windows of w columns each. Because of the use of the five point stencil, values for all points in a given window of a block may be computed before any work on the next window is begun. If one numbers the windows in each block from left to right, block i may commence work on window j when block $i-1$ has finished work on window j. This leads to a pattern of computation [15] in which a wavefront of computation is seen to propagate from the lower left portion of the domain (Figure 1). The block size and the size chosen for the window both determine the coarseness of the computation's granularity. In [15] is found a quantitative analysis of this tradeoff in the case where the block size is equal to the window size and the grid is square. This analysis is extended both through analytically and experimentally in the following in order to explore the effects of independently varying block and window size in a rectangular grid.

For a grid whose points are connected by an arbitrary stencil, the definition of work schedules that maintain data dependency relations yet allow for varying degrees of granularity is somewhat more subtle. Work is begun in the first row of the first block, and in this row the values for a window of w grid points are calculated. Following this, values are found for all mesh points in the block for which data dependencies allow calculation. The computation proceeds after this in stages, with the computations that may proceed in a block at a given time being determined by dataflow considerations. If one wishes to aggregate points in blocks into larger units, with each unit to be calculated sequentially, the partitioning will take on a zig-zag form. Figure 2 depicts the pattern of wavefronts that results from partitioning a domain with a nine point stencil into blocks of size two, and scheduling computation using a window size of two.

2.2 Automated Problem Partitioning

In order to automate problem partitioning and work scheduling, it is essential to be able to dispense with as much application dependent information as possible. We have developed and tested a method for generating a work partition in problems possessing data dependencies given by a directed acyclic graph (DAG). This method bears a strong relationship to methods proposed for systolic array generation [1], [10]. The order in which variables, described by rows in L,

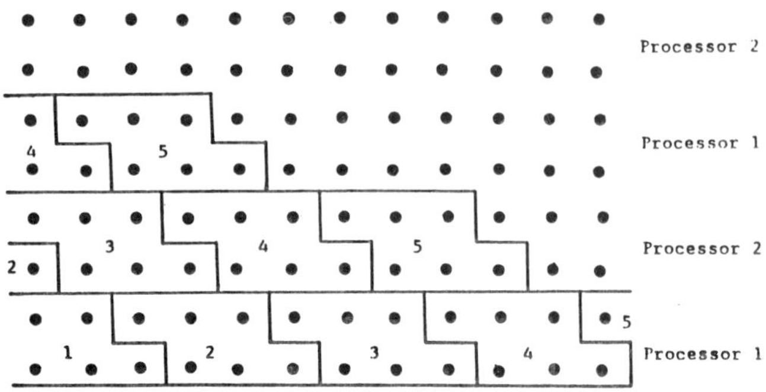

Figure 2: Two processors, nine point stencil, block size = 2, window = 3. Numbers designate computational phases.

can be solved may be depicted by a directed acyclic graph D. The evaluation of rows in L are represented by the vertices of D, and the data dependencies between the rows by D's edges. The dependence of matrix row a on matrix row b is represented by an edge going from vertex b to vertex a. A topological sort may be performed which partitions the DAG into wavefronts. A stage of this sort is performed by alternately removing all vertices that are not pointed to by edges, and then removing all edges that emanated from the removed vertices. All vertices removed during a given stage constitute a wavefront; the wavefronts are numbered by consecutive integers. An adaptation of a common topological sort algorithm [9] allows the wavefronts of a DAG to be calculated efficiently.

The wavefronts calculated through this process can be utilized directly in implementing a very general method for scheduling the row substitutions required for the solution of the equations. The row substitutions in any wavefront may be executed simultaneously. A very straightforward method for solving the problem is consequently to partition the problem's solution into phases, each of which is dedicated to a given wavefront. On shared memory machines, the straightforward application of this technique requires a global synchronization between phases. Because in many cases there is only a relatively modest amount of computation required for a given phase, the relative cost of the global synchronization can be relatively substantial, as will be shown in the experimental results below. On many message passing machines, (e.g. the Intel iPSC [13]), the communication latency makes this kind of medium grained parallelism particularly prohibitive. Similarly, in message passing machines, it is of considerable importance to map problems in a way that reduces interprocessor communication requirements.

For many problems possessing relatively regular patterns of data dependency, one can obtain a variety of benefits on both shared memory and message passing machines by carrying the run time analysis a step further through partitioning the DAG in a particular way. The points of a DAG are partitioned into disjoint sets called *strings*. A string partition of a problem is generated through the following sequence of depth first traversals in DAG D.

We define a start vertex of D as a vertex not pointed to by any edge. The vertices making up a string S are chosen in the following way. A start vertex V of D is chosen, all edges emanating from V are removed; if a new start vertex V' is created through the removal of edges, V' is included in the string. The process is continued to recursively remove as many vertices as possible from D, and assign them to S. Note that when, during the creation of string S, the removal of a vertex

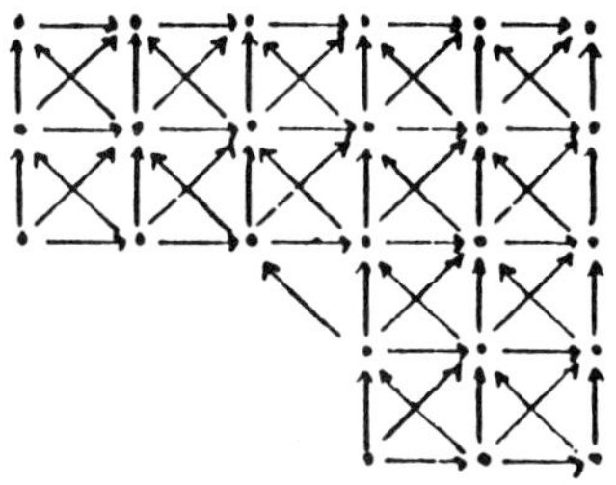

Figure 3: Data Dependencies

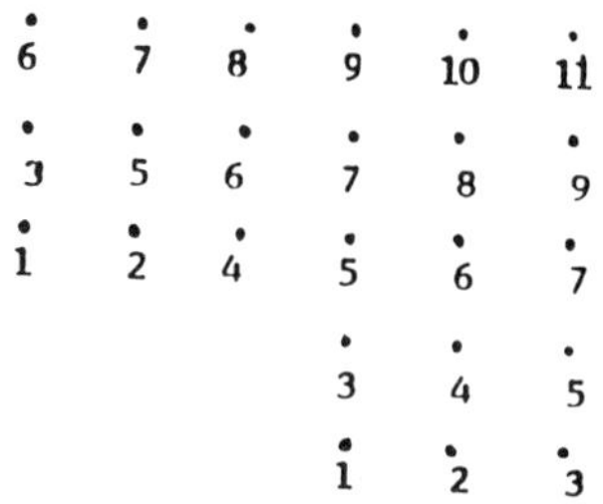

Figure 4: Wavefronts

exposes multiple start vertices, only one of these start vertices are included in S. As vertices V' are assigned to S, we mark the vertices W remaining in D that had edges arising from V'. New strings are begun using available start vertices. In choosing vertices to incorporate in all strings after the first, preference is given to vertices previously marked by other strings.

Strings have the following properties: (1) The points in each string are connected, (2) There is no more than one point belonging to a given wavefront in a string, (3) The graph describing the *inter-string* dependencies is a directed acyclic graph. The DAG describing the inter-string dependencies will be called the *string DAG.*

Figure 3 depicts a DAG which could be obtained from a zero fill incomplete factorization of a matrix arising from the discretization of an elliptic partial differential equation using a nine point star template. Figure 4 depicts the wavefronts in the computation, and figure 5 depicts a string decomposition and illustrates the string DAG corresponding to this problem.

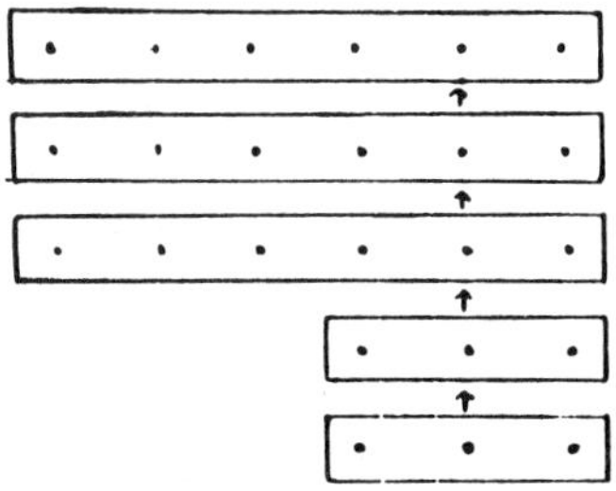

Figure 5: Strings and String DAG

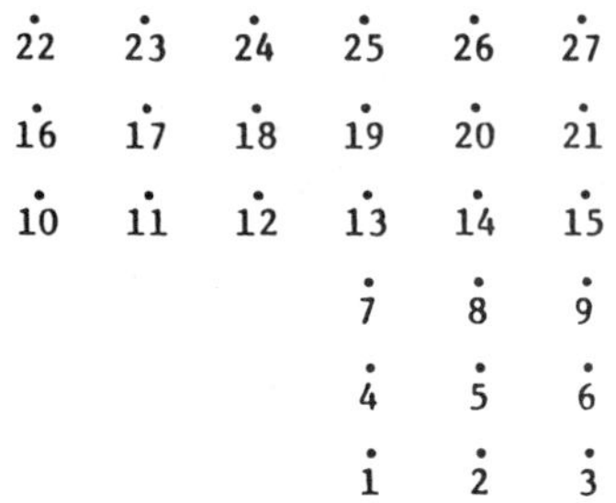

Figure 6: Mesh Point Ordering

It should be noted that it may be possible to partition a DAG into strings in several different ways. For example, the triangular system arising from the zero fill factorization of the matrix generated by a rectangular grid with a 9 point template can be partitioned in two ways. In one partitioning, matrix rows originating from horizontal strips of domain form strings, in the other matrix rows originating from diagonal strips of domain form strings. Performance implications of these different methods of decomposition are discussed in the presentation of experimental timings in [12].

When a DAG originates from the incomplete decomposition of a two dimensional domain, it is possible to make sure that the strings are chosen in a much more controlled manner. The decomposition can be determined by the way in which the mesh points are ordered in the formation of the matrix. It is simple to arrange for the algorithm to give preference to lower numbered rows when forming new strings from start vertices D, and to attempt to incorporate rows into a growing string in order of increasing row number. For example, figure 6 depicts a simple way of numbering the grid points from figures 3 and 4 to ensure the production of the string decomposition depicted in 5.

In a rough sense, the strings of a DAG D are sets of points orthogonal to the wavefronts of D. This type of decomposition allows for considerable flexibility in determining the granularity of parallelism, as discussed below. The decomposition of the DAG D into strings will be shown below to facilitate particularly inexpensive forms of synchronization in shared memory architectures.

The data dependency relationships in the problems discussed in this paper are quite regular and are easily handled by the mechanism described above and in fact could be handled by methods described in [1] if the data dependencies were given in a symbolic fashion.

2.3 Mapping Strings onto Processors

The string DAG may be distributed among processors in a variety of ways. On message passing machines, mapping large contiguous sections of the string DAG onto each processor will tend to minimize communication costs, but will also tend to lead to poor load distributions. Scattering or wrapping strings that are contiguous in the DAG may lead to a much better load distribution at the price of increased communication costs.

The work associated with each cluster of strings may be scheduled with varying degrees of granularity. The string DAG defines a partial ordering among the strings. The starting strings may be defined as the strings that precede all others in this partial ordering. Computations of rows in these strings are not dependent on information from any other strings in the string DAG.

The partial ordering of the data dependencies between the strings allows for the straightforward implementation of dataflow synchronization methods. The granularity of parallelism may be determined by fixing the amount of work starting strings can perform before communicating their data to other strings in the string DAG. Simple relationships involving the wavefronts of *rows* allows the calculation of which rows may be solved for by a processor assigned to a cluster of strings.

3 Construction of Work Schedules

3.1 Overview

The parallelism involved in solving a sparse triangular system of equations is inherently rather fine grained; the work required to compute the value of a variable corresponding to a given row generally amounts to only a few floating point operations. In scientific applications, one frequently wishes to obtain multiple solutions with either the same triangular system or triangular systems with the same non zero structure. In these cases, it seems to be appropriate to spend a modest amount of time to calculate a work schedule. It is essential, however, that very little time be required to schedule row executions during the execution of the algorithm. We consequently have chosen to use a prescheduled approach in which the rows to be computed by a string at a given phase in a computation are chosen implicitly by determining the wavefronts that should be computed during that phase.

We will now consider methods for scheduling the execution of work given a string DAG. We will assume that the strings making up the string DAG have been linearly ordered and that contiguous blocks of b strings are demarcated, and are assigned to consecutive processors in a wrapped manner.

For barrier and dataflow synchronization methods, we will present expressions denoting the largest wavefront that the strings in a block must compute during a particular phase. The expressions presented here are derived in [12]. Obviously computations cannot be undertaken until the required data is available. The calculation of the wavefronts that are to be computed during each phase takes into account the data that is guaranteed to be available when a processor reaches a given phase.

The phase during which one can assure data availability for a given computation depends on (1) the synchronization mechanism used, (2) the data dependency relationships between the strings of the string DAG and (3) the data dependencies between the blocks into which the string DAG is partitioned.

We will now discuss scheduling when barrier synchronization methods are employed; these methods assure that all processors have finished phase $p-1$ before any processor is allowed to begin phase p.

The proposition below presents expressions that give the maximum wavefront number that is to be computed by a given block i during phase p, under the assumption that the first block computes exactly w wavefronts per phase; i.e. during phase p the first block computes wavefronts $w(p-1)+1$ to wp.

This proposition may be regarded as a method of parametrically describing the wavefronts of a coarse grained DAG, each vertex of which represents the solution of a number of rows (note: this coarse grained DAG is *not* the string DAG). A wavefront of this coarse grained DAG will be called a *block wavefront*. It should be noted that one may assign any work scheduled during a phase to any processor one desires. In problems described by irregular DAGs it will be shown below that explicitly balancing the processor load in a block wavefront during each phase can be quite advantageous.

Proposition 1 *Assume that strings making up the string DAG have been linearly ordered, that contiguous blocks of strings are demarcated, and that these blocks are assigned to consecutive*

processors in a wrapped manner. Let W_p^i represent the largest wavefront that can be scheduled during phase p by block i under the following conditions: (1) the first block advances w wavefronts per phase, i.e. $W_p^1 = wp$, and (2) all required data is computed before the system reaches phase p.
W_p^i is given by the expression $W_p^i = \max(p, w(p - i + 1) + i - 1)$.

When it is known that data dependencies occur only between adjacent strings, a more aggressive scheduling policy can be used.

Proposition 2 *Assume that strings making up the string DAG have been linearly ordered so that data dependencies occur only between adjacent strings, that contiguous blocks of strings are demarcated, and that these blocks are assigned to consecutive processors in a wrapped manner. Let W_p^i represent the largest wavefront that can be scheduled during phase p by block i under the following conditions: (1) the first block advances w wavefronts per phase, i.e. $W_p^1 = wp + b - 1$, and (2) all required data is computed before the system reaches phase p.*
W_p^i is given by the expression

$$W_p^i = \begin{cases} w(p - i + 1) + ib - 1 & \text{if } p \geq i \\ bp & \text{if } 0 \leq p < i. \end{cases}$$

When dataflow synchronization is used, at any given time, processors may be performing computations specified by different phases. To ensure that needed data is available when a processor performs its computations during a phase, we must construct a schedule that makes adequate allowance for the weak interprocessor synchronization. The calculation of the wavefronts that are to be computed during each phase takes into account the data that is guaranteed to be available when a processor reaches a given phase. The schedule to be presented below also takes into account the data dependencies between blocks of strings in the string DAG.

To illustrate the role of inter-block data dependencies in determining schedules for performing work when dataflow synchronization is used, consider an important special case that arises when inter-block data dependencies are restricted to nearest neighbors, i.e. block $i + 1$ requires data only from block i. In this case the constraints for scheduling wavefront execution during a given phase p are identical for dataflow and barrier synchronization mechanisms. In either of these synchronization mechanisms, a processor P beginning phase $p + 1$ has to know that its predecessor has finished phase p. Block i assigned to P requires data only from block $i - 1$ so that no provision need be made for the possibility that blocks upon which i depends have not yet computed values for wavefronts corresponding to phase p.

When dataflow synchronization is employed, proposition 3 below presents expressions that give the maximum wavefront number that is to be computed by a given block i during phase p, under the assumption that (1) the first block computes exactly w wavefronts per phase, and (2) block i can require data only from blocks j, $\max(i - d, 1) \leq j < i$. Note that when $d = 1$ we have the previously discussed case of nearest neighbor data dependencies. The string DAGs that arise from many problems obtained from incomplete factorizations of matrices arising from partial differential equations are frequently characterized by small values of d.

The expression is obtained by mapping a chain of logical processes in a wrapped manner onto the P processors of the machine. For the sake of tractability, the expressions are derived under the assumption that the logical processes assigned to a given processor are assumed to be independent of one another. In the actual system, all processes assigned to the same processor must complete a given phase before beginning the next. Creating schedules using the assumption that processes assigned to a processor are independent assures us that computations will not be undertaken until the required data are available, since such a schedule allows for the possibility that all processes on a given processor could be computing the same phase at a given time. When

$d < P$, the expressions derived yield the largest wavefront that could be scheduled by a given block at a particular time, when $d \geq P$ this is no longer necessarily the case.

Proposition 3 *Assume that strings making up the string DAG have been linearly ordered, that contiguous blocks of strings are demarcated, and that these blocks are assigned to consecutive processors in a wrapped manner. Assume that each block constitutes a process that executes its computations in phases subject to the constraint that at any time, if block i has finished phase p, block $i+1$ can complete all phases with numbers less than or equal to $p+1$. Furthermore assume that each block i requires data only from blocks* $\max(i-d,1)$ *through* $i-1$.

Let W_p^i represent the largest wavefront that can be scheduled during phase p by block i under the following conditions: (1) the first block advances w wavefronts per phase, i.e. $W_p^1 = wp$, and (2) all required data is computed before the system reaches phase p.

For $i \geq 2$, W_p^i is given by the expression

$$W_p^i = \max(\lceil p/d \rceil, w(p-i+1) + \lceil (i-1)/d \rceil) \tag{1}$$

4 Load Balance - Synchronization Cost Tradeoffs

4.1 Analysis of a Model Problem

For a given problem, the tradeoffs between load imbalance and synchronization costs will vary with choice of window and block size. We will examine this tradeoff in the context of solving a lower triangular system generated by the zero fill factorization of the matrix arising from a rectangular mesh with a five point template. We will utilize P processors and partition the domain into n horizontal strips where each strip is divided into m blocks, as is depicted in figure 1. We will assume that the problem is obtained from a domain with N by M mesh points, and that all computations required to solve the problem would require time S on a single processor. We will also assume that computation of each block takes time $T_B = S/(mn)$; this ignores the relatively minor disparities caused by the matrix rows represented by points on the lower and the left boundary of the domain. Horizontal strips of blocks are assigned to each of P processors in a wrapped manner. The computation is divided into phases; during phase p the processor assigned to strip i computes block $p-i+1$ in the strip, as long as $1 \leq p-i+1 \leq n$.

A brief inspection of figure 1 makes it clear that $n+m-1$ phases required to complete the computation. Define $MC(j)$ as the maximum number of blocks computed by any processor during phase j. The computation time required to complete phase j is equal to $T_B MC(j)$, the computation time required to complete the problem is consequently

$$\sum_{j=1}^{n+m-1} T_B MC(j).$$

We now proceed to calculate $MC(j)$. During phase j, a total of j blocks must be computed when $1 \leq j < \min(m,n)$. Since the blocks are assigned in a wrapped manner,

$$MC(j) = \lceil \frac{j}{P} \rceil.$$

When $\min(m,n) \leq j \leq n+m-\min(m,n)$, a total of $\min(m,n)$ blocks must be completed during phase j. Due to the wrapped assignment of blocks to processors,

$$MC(j) = \lceil \frac{\min(m,n)}{P} \rceil.$$

Finally when $n + m - \min(m,n) < j \leq n + m - 1$, a total of $n + m - j$ blocks must be computed during phase j so

$$MC(j) = \lceil \frac{n+m-j}{P} \rceil.$$

The computation time required to complete the problem is consequently

$$T_B \sum_{j=1}^{n+m-1} MC(j) =$$

$$\frac{S}{mn}\Big(\sum_{j=1}^{\min(m,n)-1} \lceil \frac{j}{P} \rceil + (n + m - 2\min(m,n) + 1) \lceil \frac{\min(m,n)}{P} \rceil +$$

$$\sum_{j=m+n-\min(m,n)+1}^{n+m-1} \lceil \frac{n+m-j}{P} \rceil \Big)$$

In a shared memory environment we must synchronize between phases. Assume that each synchronization has cost T_S. The total time spent synchronizing is then given simply by $T_S(n+m-1)$. Assume the problem is mapped to a message passing machine so that processors assigned consecutive strips of blocks directly communicate and where links between processors can operate in parallel. The cost of sending a B word message between two processors can be approximated as $\alpha + \beta B$. We will make the further approximation concerning the cost of requiring *each* processor to send a message to its neighbor following phase j. That cost is equal to the time required to communicate the largest message sent between two processors following phase j. The maximum amount of information that must be sent from one processor to another after phase j is $(M/n)MC(j)$. The cost of communications that follow phase j may be expressed as

$$\alpha + \beta \frac{M}{n} MC(j).$$

The total cost of communications is hence given by

$$\alpha(n + m - 1) + \beta \frac{M}{n}$$

$$\left(\sum_{j=1}^{n+m-1} \lceil \frac{j}{P} \rceil + (n + m - 2\min(m,n) + 1) \lceil \frac{\min(m,n)}{P} \rceil + \right.$$

$$\left. \sum_{j=m+n-\min(m,n)+1}^{n+m-1} \lceil \frac{n+m-j}{P} \rceil \right)$$

Using the above considerations, we will now calculate a simple expression for the amount of work forgone during a computation when the number of blocks in a strip n as well as the number of strips in a problem m are both integer multiples of the number of processors P utilized. We thus assume that $n = r_1 P$ and $m = r_2 P$, for r_1, r_2 positive integers. From the discussion above, during the first $\min(m,n) - 1$ phases, the computation requires time

$$T_B \sum_{j=1}^{\min(m,n)-1} \lceil \frac{j}{P} \rceil.$$

During phase $j \leq \min(m,n) - 1$ when j is not a multiple of P, there are $P - j \bmod P$ processors idle; when j is a multiple of P, no processors are idle. Thus the sum of the processor idle time for $j \leq \min(m,n) - 1$ is $T_B \ \min(r_1, r_2) \sum_{l=1}^{P} (l - 1)$, or

$$T_B \frac{\min(r_1, r_2) P(P-1)}{2}.$$

Through identical arguments, the sum of the processor idle time for the last $\min(m,n)-1$ phases is the same as that above. During the intermediate phases the load is balanced with $\min(m,n)$ blocks assigned to each processor.

In a shared memory environment, using the above expressions for the processor time wasted due to load imbalance and the time spent in synchronization we find that the total processor time wasted from both causes may be given by:

$$\frac{SP(P-1)\min(r_1,r_2)}{r_1 r_2 P^2} + T_S P(r_1 P + r_2 P)$$

Note that the above expression is symmetric with respect to r_1 and r_2. When the synchronization cost is the dominant overhead, it consequently does not matter whether one uses small windows and assigns large blocks of variables to each processor, or whether one uses large windows and assigns small blocks of variables to each processor. Assume without loss of generality that $r_2 \leq r_1$, i.e. that the window size is at least as large as the number of rows of grid points in a block. In this case the total processor time wasted is

$$\frac{SP(P-1)}{r_1 P^2} + T_S P(r_1 P + r_2 P). \tag{2}$$

For any fixed r_1 reducing r_2 to 1 decreases the time spent by processors in synchronization without impacting adversely on the balance of computational load. Thus when the number of blocks in a strip n as well as the number of strips in a problem m are both integer multiples of the number of processors P utilized, the window size can be profitably increased to M/P. This increase in window size does not affect the distribution of load and reduces the number of phases required to solve a problem.

5 Wavefront Longest Processing Time Scheduling

Propositions (1) and (2) describe a parametric method of constructing work schedules for performing the calculations required to solve the problem in question, when a form of barrier synchronization is used between phases. As stated previously, these propositions may also be regarded as a way of parametrically describing the wavefronts of a coarse grained DAG, each vertex of which represents the solution of a number of rows. It is consequently natural to consider balancing the processor load for the block wavefronts of this DAG, i.e. balancing the load during each phase of computation. We choose to utilize a prescheduled approach for allocating work although it is also possible to allocate work represented in these wavefronts in a self scheduled manner.

The scheduling of independent tasks to obtain a minimum finishing time is known to be NP-hard. There exist a variety of methods for obtaining approximate solutions to this problem [5], [8]. One method that has been extensively studied is the *Longest Processing Time* or the *LPT* schedule. An LPT schedule is one that is the result of an algorithm which, whenever a processor becomes free, assigns to that processor a task which requires a run time longer than that of any task not yet assigned.

While the relative difference in performance between results obtained with LPT and optimal schedules in the worst case is $1/3 - 1/(3P)$ where P is as usual the number of processors [5] , in simulations investigating the error that might be expected given a variety of randomly generated datasets, it was found that the difference between the generated solution and the optimal solution was only a few percent [2]. Because the prescheduled work assignment must use approximate work estimates based on calculations involving the number of floating point operations required to solve for a row of the triangular system, a heuristic such as LPT is likely to perform as well as a more expensive scheme to find a closer approximation to the optimal schedule.

The LPT rule requires time $r \log r$ to schedule the execution of r tasks. For a fixed choice of window w and block size b in the work schedule, the amount of computation in a wavefront of a triangular solve arising from a zero fill incomplete factorization of a matrix generated by a regular n by n mesh increases with order n. In an asymptotic sense then, even the rather inexpensive LPT scheduling algorithm has somewhat unfavorable properties. The performance obtained through the use of the LPT scheduling algorithm is compared with that obtained through the use of a wrapped assignment of strings in the following section.

6 Experimental Results

6.1 Preliminaries

The figures discussed in the current section depict the results of measurements made of the amount of time required to perform a forward substitution utilizing the three forms of synchronization discussed above. The matrix utilized was generated through the zero fill incomplete factorization of square meshes of various sizes, in which one of a number of templates were employed.

Before discussing the experimental results in detail, the architecture of the Encore Multimax will be briefly described. The Encore Multimax is a bus based shared memory machine that utilizes 10 MHz NS32032 processors and NS32081 floating point coprocessors. Processors, shared memory, and i/o interfaces communicate using a 12.5 MHz bus with separate 64 bit data paths and 32 bit address paths.

Associated with each pair of processors is a 32K-byte cache of fast static RAM. Memory data is stored in this cache whenever either of the two processors associated with the cache reads or writes to main memory locations. Each cache is kept current with the relevant changes in main memory by continuous monitoring of traffic on the bus [3].

All tests reported were performed on a configuration with 16 processors and 16 Mbytes memory at times when the only active processes were due to the author and to the operating system. On the Encore the user has no direct control over processor allocation. Tests were performed by spawning a fixed number of processes and keeping the processes in existence for the length of each computation. This programming methodology is further described in [11]. The processes spawned are scheduled by the operating system, and for this otherwise empty system, throughout the following discussions we make the tacit assumption that there is a processor available at all times to execute each process. In order to reduce the effect of system overhead on our timings, tests were performed using no more than 14 processes; this left two processors available to handle the intermittent resource demands presented by processes generated by the operating system.

It should be noted that the bus connecting processors to memory does not appear to to cause significant performance degradation in problems with the mix of computations and memory references that characterize the problems described here. In a set of experiments using a variety of sparse lower triangular matrices, multiple identical sequential forward solves were run on separate processors at the same time. Timings of this experiment exhibited performance degradations of less than one percent as one increased the number of processors utilized from one to 14.

6.2 Effect of Window and Block Size on Performance

The effect of window size on execution time was investigated. The data depicted in figure 7 was obtained through a forward solve of the zero fill factorization of a matrix generated using a 100 by 100 point square mesh, in which a 5 point template was employed. Barriers were used for synchronization. Note that this matrix is extremely sparse, there are no more than two non-zero off diagonal elements in any matrix row. This matrix, along with the others described

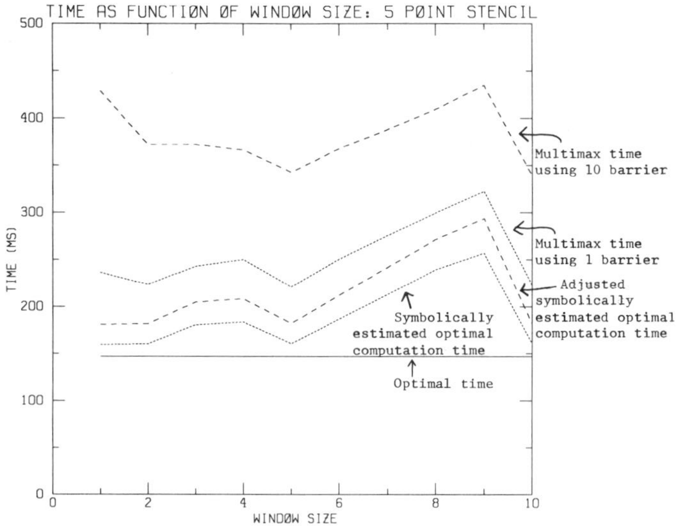

Figure 7: Effect of Window Size on Execution Time. Matrix from a 100 by 100 mesh, 5 point template. 10 processors used, timings for 25 consecutive trials averaged.

here, has unit diagonal elements. The forward solve consequently does not involve divisions. The parallelism encountered here is consequently quite fine grained. The strings in this problem partition the domain into horizontal slices as was described in the analysis of the model problem previously discussed. Horizontally oriented strings were used in all experimental results reported here; a discussion of performance effects of string orientation may be found in [12]. Ten processors were used to solve this problem, and a block size of one was employed.

A symbolic estimate was made of the optimal speedup that could be obtained in the absence of synchronization delays, given the assignment of work to processors characterizing a particular window and block size. For each window size, the time required for a separate sequential code to solve the problem was divided by the estimated optimal speedup. This yields the amount of time that would be required to solve the problem in the absence of any sources of inefficiency other than load imbalance. The results of these calculations are plotted in figure 7 where they are denoted as the symbolically estimated optimal computation time.

Dividing the execution time of the one processor version of the parallel code by the estimated optimal speedup yields a further refined estimate of the shortest amount of time in which the problem could be solved in the absence of synchronization delays. In figure 7 these results are plotted and are denoted as the adjusted symbolically estimated optimal computation time. It is of interest to note that, as predicted by equation (2), the two estimates of the optimal computation time predict close to identical computation times for windows of size one, two, five and ten. The computation times estimated for windows of other sizes are larger.

Timings were obtained by solving the problem using ten processors on the Multimax, timings were averaged over 25 consecutive runs. Barrier synchronization between phases was utilized. When timed separately, this synchronization was found to require 75 microseconds; this compares to approximately 20 microseconds required for a single precision floating point multiply and add. It is not clear that future architectures utilizing much faster processors and more general interconnection networks will allow for synchronization costs that are as small relative to the costs of floating point computation. In a separate set of measurements also depicted in figure 7, the effects of varying window size in an environment characterized by higher relative synchronization costs were explored though the use of ten 75 microsecond barriers between phases of the computation.

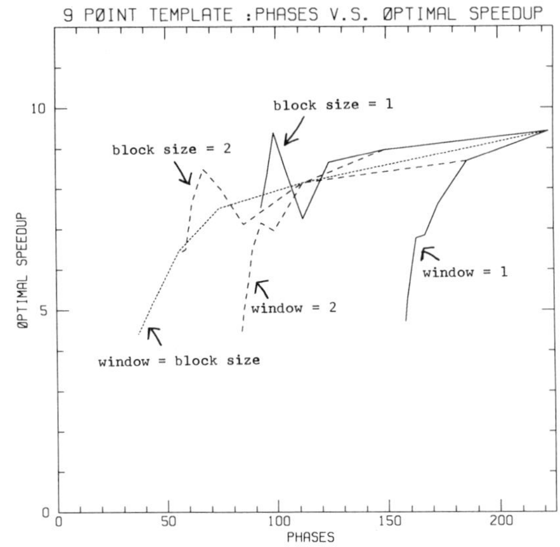

Figure 8: Symbolically estimated optimal speedup versus phases required to solve problem on 12 processors. Matrix from a 75 by 75 point mesh, 9 point template.

Finally, the time required to solve the the problem using the sequential code was divided by the number of processors used; figure 7 this is depicted, and is denoted there as optimal time.

Tradeoffs between load imbalance and synchronization costs were examined in a different manner by comparing the symbolically estimated optimal speedup against the number of phases required to complete a problem. The symbolically estimated optimal speedup takes into account the degree to which a given assignment of work to processors balances the workload. The number of phases required to solve a problem has a strong bearing on the synchronization overhead encountered in solving a problem.

In figure 8 the symbolically estimated optimal speedup was compared with the phases required for solving a lower triangular system generated by zero fill factorization of a matrix arising from a 75 by 75 point mesh, utilizing a nine point template. The strings chosen were those partitioning the domain into horizontal strips.

The estimated speedups resulting from the use of blocks of sizes one and two, with the size of windows varying from one to eight are depicted, along with the speedups resulting from the use of windows of sizes one and two, with the size of blocks varying from one to eight. Also depicted are speedups resulting from using a block size that is equal to the window size; both are varied from one to six.

The tradeoff between speedup and number of phases used, appears to be generally more advantageous when large windows and small block sizes are used than when the situation is reversed. As was observed in the examination of the performance obtained using the five point template, the number of phases declines with increasing window and or block size, while the load balance exhibits substantial fluctuations. The tradeoff between load imbalance and number of phases required appears to be much smoother when the size of the window used is set equal to the size of the block than in the other cases discussed above; the estimated speedup is in this case a decreasing function of the block and window size used. For any given number of phases, the load balance when window size is equal to block size is superior to that obtained when the window size is set equal to one or two and the block size is varied.

As synchronization costs increase, it becomes more advantageous to reduce the number of

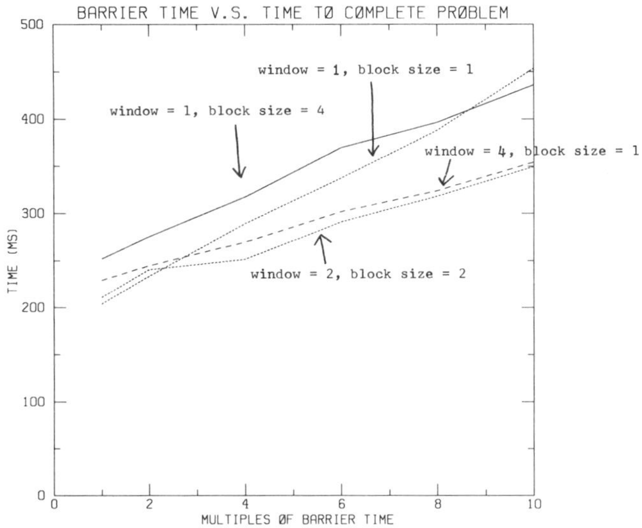

Figure 9: Effect of window, block size on execution time. Matrix from a 75 by 75 point mesh, 9 point template. 12 processors used, timings for 25 consecutive trials averaged.

phases required to solve a problem even at the cost of increased load imbalances.

The relative performance of four combinations of window and block size in the face of increasing synchronization costs are depicted in figure 9. The execution time required to solve the lower triangular system described above was measured when the following combinations of window and block size were employed: (1) window size = 1, block size = 1, (2) window size = 4, block size = 1, (3) window size = 1, block size = 4 and (4) window size = 2 , block size = 2. Between phases, we employed from one to ten 75 microsecond barrier synchronizations.

The numbers of phases and the symbolically estimated optimal speedup for each of these cases are listed below.

block size	window	phases	est. speedup
1	1	223	9.43
1	4	112	7.26
4	1	167	6.84
2	2	112	8.14

As one can observe from the above table, block size four, window one and block size one, window four in this problem require at least as many phases as does block size two, window two and the later achieves a superior load balance. The use of block size one, window one allows one to achieve a load balance that is even better, but at the cost of added phases of computation. In figure 9, for barrier times between 75 and 150 microseconds, the shortest run times were obtained using block size and window size both equal to one. When barriers were utilized that required more than 150 microseconds the use of block and window sizes both equal to two lead to the shortest run times.

6.3 Comparison Between LPT and Wrapped Scheduling

When barrier synchronization is utilized, the work required to compute the blocks during a phase can be scheduled in an explicit manner. We have discussed how one might use an inexpensive heuristic such as LPT to do this scheduling. Experimental comparisons will now be made between

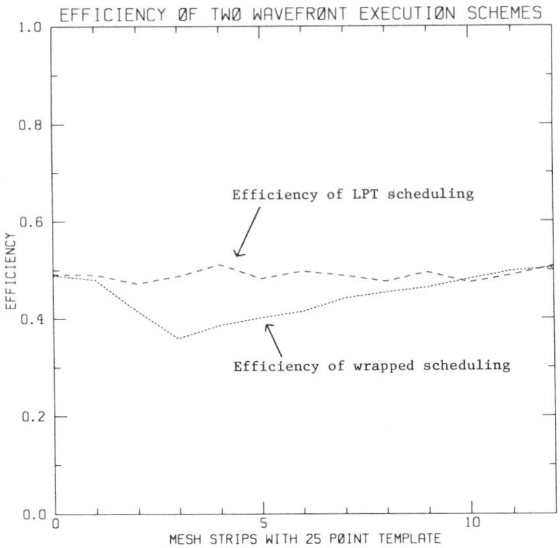

Figure 10: Efficiency of wrapped vs LPT scheduling. From Multimax times. Matrix from 100 by 40 mesh, bottom strips have 25 point template, rest have 5 point template. 12 processors, block,window equal 1, timings for 25 trials averaged.

the performance that can be achieved through the use of LPT and that obtained by assigning the workload to the processors in a wrapped fashion.

The difference in performance between these scheduling methods is only expected to be noticeable in problems with some degree of irregularity. If during each phase a number of blocks with identical computational requirements had to be executed, a wrapped assignment should lead to an optimal balance of load during that phase. Note that this does not mean that the load will be balanced, since the number of blocks assigned to processors can differ by one.

Of course, the computational requirements of blocks to be computed during a phase are not identical, even in problems derived from rectangular meshes in which a uniform template was utilized. In such problems, the blocks derived from mesh points near the boundaries of the domain will generally require smaller amounts of computation than those derived from points further away from the boundaries. Two sets of experiments were performed to compare LPT and wrapped scheduling using a matrix generated from a 80 by 80 point mesh with a 5 point template and a matrix generated from the same mesh using a 13 point template. Block and window size were varied and 10 processors were used. The performance obtained using the two scheduling methods were compared using both symbolically estimated optimal speedups and measured runtimes on the Encore Multimax. The symbolically estimated optimal speedups were not effected by the scheduling mechanism used, and the Multimax runtimes measured showed minimal differences in no consistent direction.

A set of problems exhibiting more dramatic load imbalances were examined. A lower triangular system was produced by the incomplete factorization of a matrix generated from a 100 by 40 mesh point matrix in which the bottom s strips had a 25 point template, and the $40 - s$ upper strips had a 5 point template. Both the block size and the window size were set equal to one and a single barrier was used for synchronization. Figure 10 depicts the efficiency with which 12 processors of the Multimax solves the system as s is varied from 0 to 12. Efficiency is defined here as the ratio of the time required to solve the problem using a separate sequential code on one processor to the product of the measured time to solve the problem and the number of processors used.

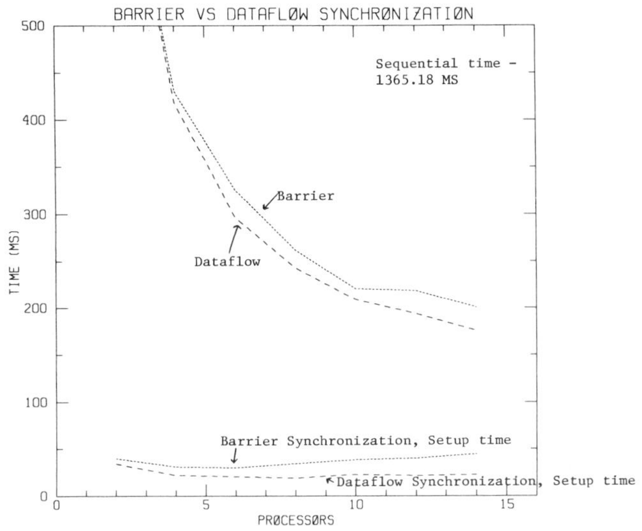

Figure 11: Run time of wrapped and LPT scheduling. Matrix from 100 by 40 mesh, bottom 5 strips have 25 point template, other strips have 5 point template. 12 processors, block, window equal 1. Timings for 25 trials averaged.

The efficiency obtained through the use of LPT scheduling does not vary much with s, remaining approximately 0.50. The efficiency exhibited by the wrapped scheduling decreases to a low of 0.36 for s equal to 3, but is comparable to the efficiency obtained through the use of LPT when s is close to either 0 or 12. The reasons for this appear to be quite straightforward. When one has, during each phase, a number of very time consuming blocks that is small compared to the number of processors used, one risks a serious load imbalance when a wrapped assignment strategy is used. As the number of time consuming blocks encountered during each phase increases to approach the number of processors, the amount of wasted processor capacity decreases.

Further insight into the situation is provided by figure 11, in which the problem described above is solved when $s = 5$ for varying numbers of processors. In this figure we display the symbolically estimated optimal computation times (abbreviated as SEOC in the figure) obtained from both the wrapped and the LPT schedules, along with the measured Multimax execution times. As in previous figures, the optimal time is defined as the sequential time divided by the number of processors. The time required for both synchronization and for executing the program control structure was also measured by making program measurements with the floating point calculations commented out.

Insight into the sources of inefficiency in this problem may be obtained by examining this figure. There is a substantial difference between the optimal time and both symbolically estimated optimal computation times, suggesting that load imbalance makes a significant contribution to the departure from the optimal run time observed here. This conclusion is reinforced by observing the measured synchronization and setup time.

It may be noted in this figure that the symbolically estimated optimal computation time calculated for the wrapped assignment added to the synchronization and setup time produce numbers that are quite close to the measured multimax time for wrapped scheduling. This correspondence has been noted in the case of the wrapped assignment in a variety of other problems not presented here and gives confidence in the accuracy of the measurements. When LPT scheduling is used, the symbolically estimated optimal computation time appears have less predictive value, this may be seen from figure 11, and has been noted in other measurements.

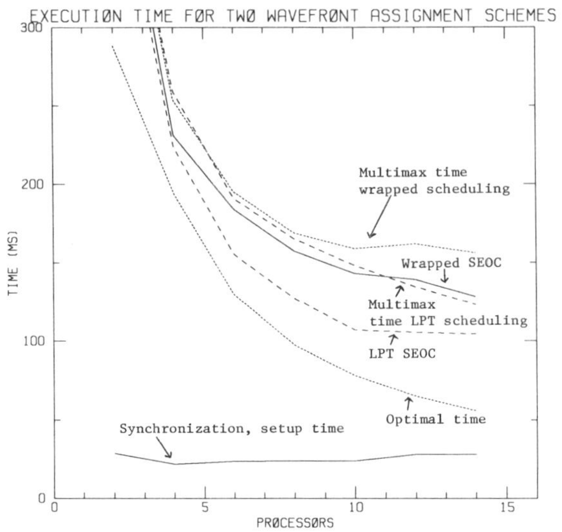

Figure 12: Execution time of barrier and dataflow synchronization. Matrix from 75 by 75 mesh, 9 point template. Window, block equal 1. Timings from 25 consecutive trials averaged.

For instance, while the efficiencies obtained through using LPT in figure 10 vary little with s, the symbolically estimated optimal computation times vary with s to a substantial extent. For instance for s equal to 2, the measured time taken to solve the problem using the LPT algorithm was 112.16, while the SEOC was 69.4; for s equal to 3, the measured time was 118.3 but the SEOC was 99.31. The difference in synchronization time between the two cases was, however, quite minimal.

It should be remembered however, that the LPT scheduling method uses symbolically estimated computation times to perform its load balancing. SEOC estimates are clearly not completely accurate. If one measures the SEOC obtained, after rescheduling computations using a method that produces a near optimal processor schedule based on operation counts, one will tend to obtain overly optimistic estimates of the execution time. This observation points to the obvious importance of accurate run time estimates when performing LPT scheduling.

6.4 A Comparison between Barrier and Dataflow Synchronization

While barrier synchronization is relatively inexpensive on the Encore Multimax, nearest neighbor synchronization is less expensive still as it can be implemented in a way that requires, each processor, only one shared variable increment followed by a busy wait. Figure 12 depicts a comparison between execution times measured when a barrier was utilized and execution time measured when dataflow synchronization was employed. Both barrier and dataflow synchronization - setup time are also measured and depicted. The problem solved here originates from a 75 by 75 point mesh with a 9 point template, the window and block size are 1. It is evident from this figure that dataflow synchronization is less expensive than barrier synchronization.

7 Conclusion

We have carried out an investigation into methods appropriate for the aggregation, mapping and scheduling of relatively fine grained computations specified by a directed acyclic graph. The solution of very sparse triangular linear systems provides a useful model problem for use in exploring these heuristics. A method for using the triangular matrix to generate a parameterized assignment of work to processors was described along with simple expressions that describe how to schedule computational work with varying degrees of granularity. These expressions are of considerable practical importance because they allow one to easily determine what computations need to be performed during a given phase to ensure that all required data are computed before they are required. The tradeoffs between load imbalance and synchronization costs as a function of block and window size were examined in the context of a model problem and it was demonstrated that increases in the granularity of parallelism can, in some circumstances, be obtained without any increase in load imbalances.

Experimental timings on an Encore Multimax shared memory multiprocessor confirmed the above observation in the case of the model problem and went on to explore the effects of block size and window size on multiprocessor performance in a wider variety of settings. The ratio between the costs of synchronizing and the costs of performing computations on the current Multimax is low enough that rather fine grained parallelism can be profitably used. Examination of load imbalance / synchronization cost tradeoffs in architectures requiring coarser grained parallelism was performed by experimentally varying the cost of synchronization.

As was to be expected from the model problem analysis, there is often not a particularly smooth tradeoff between between load imbalance and computational granularity. Studies of this tradeoff obtained through comparing operation counts and number of computational phases required to solve a problem suggest that the load balance granularity tradeoff becomes smoother when window size and block size are roughly equal.

The Longest Processing Time scheduling heuristic was used to explicitly schedule the work performed during each phase. The performance obtained through the use of this method was compared to the time required for assigning blocks to processors in a wrapped manner. LPT scheduling was found in some cases to decrease the run time in problems with irregular work demands during a typical phase; it had no measurable effect in very uniform problems. Through the use of symbolically estimated optimal computation times obtained through operation counts, along with direct measurements of synchronization times; for both LPT and wrapped scheduling, the origins of the measured run times were traced.

An experiment was conducted comparing the execution costs incurred through the use of barrier and dataflow synchronizations on the Multimax. Despite the use of a rather efficient barrier, time was clearly saved though the use of the even less expensive dataflow synchronization method. Dataflow synchronization has, however, a number of drawbacks. When non local data dependences occur between blocks, we see from proposition 3 that the number of phases required to complete a problem increases. Furthermore it does not appear that one can easily make use of methods for balancing the load between phases when dataflow synchronization is employed.

To sum up, in this paper we present a framework for partitioning very sparse triangular systems of linear equations that appears to be flexible enough to produce favorable performance results in a wide variety of parallel architectures. In this paper we have used the Multimax as a hardware simulator to investigate the performance effects of using the partitioning techniques presented here in shared memory architectures with varying relative synchronization costs.

A few comments are in order on which of these techniques we can recommended for use on the current Multimax. On the Encore Multimax, due to it's low ratio of synchronization costs to costs of floating point operations, there does not appear to be an advantage in aggregating work to increase the computational granularity. Balancing load within each phase of computation does appear to be advantageous in this architecture, although further practical experience is required to discover when the overhead required for this extra stage of scheduling is worthwhile. The use of dataflow synchronization on the Multimax also appears to be advantageous although it's use precludes that of wavefront LPT balancing. One can limit the problem decomposition process to the identification of wavefronts if one has no need to increase granularity through the use of windows or to use strings to implement dataflow synchronization. Hence, on the Encore there should be no reason to pay both the overhead for string decomposition and for LPT balancing.

8 Acknowledgements

I wish to thank Stan Eisenstat for making available an inexpensive barrier he has developed on the Multimax and to thank Stan Eisentat, Dave Nicol and Martin Schultz for useful discussions.

References

[1] M. C. Chen. A design methodology for synthesizing parallel algorithms and architectures. *Journal of Parallel and Distributed Computing*, 116–121, 1986.

[2] E. Coffman, M. Garey, and D. Johnson. An application of bin packing to multiprocessor scheduling. *SIAM Computing*, 7(1):1–17, 1978.

[3] *Multimax Technical Survey*. Technical Report 726-01759 Rev A, Encore Computer Corporation, 1986.

[4] A. George, M.T. Heath, J. Liu, and E. Ng. *Solution of Sparse Positive Definite Systems on a Shared-Memory Multiprocessor*. Technical Report ORNL/TM-10260, Oak Ridge National Laboratory, January 1987.

[5] R. Graham. Bounds on multiprocessor timing anomalies. *SIAM Jr. on Appl. Math*, 17(2):416–429, 1969.

[6] A. Greenbaum. *Solving Sparse Triangular Linear Systems Using Fortran with Paralllel Extensions on the NYU Ultracomputer Prototype*. Report 99, NYU Ultracomputer Note, April 1986.

[7] M. T. Heath and C. H. Romine. *Parallel Solution of Triangular Systems on Distributed Memory Multiprocessors*. Technical Report ORNL/TM-10384, Oak Ridge National Laboratory, March 1987.

[8] E. Horowitz and S. Sahni. *Fundamentals of Computer Algorithms*. Computer Science Press, Rockville Maryland, 1978.

[9] E. Horowitz and S. Sahni. *Fundamentals of Data Structures*. Computer Science Press, Rockville Maryland, 1983.

[10] Delosme J-M and Ilse Ipsen. An illustration of a methodology for the construction of efficient systolic architecture in vlsi. In *Proceedings of the Second International Symposium on VLSI Technology, Systems, and Applications*, pages 268–273, May 1985.

[11] J. F. Jordan, M. S. Benten, and N. S. Arenstorf. *Force User's Manual*. Department of Electrical and Computer Engineering 80309-0425, University of Colorado, October 1986.

[12] J. Saltz. *Automated Problem Scheduling and Reduction of Communication Delay Effects; submitted for publication*. Report 87-22, ICASE, May 1987.

[13] J.H. Saltz, V. K. Naik, and D.M. Nicol. Reduction of the effects of the communication delays in scientific algorithms on message passing mimd architectures. *SIAM J. Sci. Stat. Comput*, 8(1):s118, 1987.

[14] Joel Saltz and M.C. Chen. Automated problem mapping: the crystal runtime system. In *The Proceedings of the Hypercube Microprocessors Conf., Knoxville, TN*, September 1986.

[15] M. Schultz Y. Saad. *Parallel Implementations of Preconditioned Conjugate Gradient Methods*. Department of Computer Science YALEU/DCS/TR-425, Yale University, October 1985.

BLOCK ALGORITHMS FOR PARALLEL MACHINES

Robert Schreiber
Computer Science Department
Rensselaer Polytechnic Institute
Troy, New York 12180-3590

0. Introduction

In this paper we discuss block methods in matrix computation and the role they are beginning to play on parallel computers.

Block algorithms are advantageous for at least two important reasons. First, they work with blocks of data having b^2 elements, performing $O(b^3)$ operations. The $O(b)$ ratio of work to storage means that processing elements with an $O(b)$ ratio of computing speed to input/output bandwidth can be tolerated. Second, these algorithms are usually rich in matrix multiplication. This is an advantage because nearly every modern parallel machine is good at matrix multiplication. In fact, by attaching a systolic array to a general purpose machine, it is now easy to provide matrix multiply as a fast hardware primitive. The FPS 164-MAX is an example of a machine with this sort of capability.

Systolic arrays provide another important motivation. Future scientific machines may be equipped with systolic devices for several matrix operations – the LU, QR, and Schur decompositions, for example. There will, however, be limits, imposed by the size of the systolic array, on the order of the matrix that such a systolic functional unit can handle. A block algorithm is often just the thing for solving a matrix problem that is too big to fit on the given array.

In this regard, the situation is similar to that on the CRAY and other vector computers, which may only perform vector operations on short vectors held in vector registers. An obvious block vector algorithm (know as "strip mining") is needed to operate on longer vectors.

We distinguish in this paper between *partitioned* and *block* algorithms. A partitioned algorithm is one in which the operations of a scalar algorithms are reordered and grouped into matrix operations. If, for example, we multiply matrices A and B with the program

```
for j ← 1 to n
    for p ← 1 to n
        for i ← 1 to n
            c_ij ← a_ip b_pj.
```

we are using a point algorithm – the *middle product* method. Note that for every column of C, all of A is touched. Now suppose that $n = bk$. We may reorganize the code (or partition the algorithm) to obtain

```
for p_1 ← 1 to k
    for j ← 1 to n
        for p ← (p_1 − 1)b + 1 to p_1 b
            for i ← 1 to n
                c_ij ← a_ip b_pj
```

in which compuation of C has been broken into k passes, in any one of which only one *block column* of A and one *block row* of B need to be held in fast storage.

If bn words is to large a quantum of data, we may partition the i-loop as well. This leads to a program with five loops that requires only $b^2 + O(b)$ words of fast storage.

Dongarra and Sorensen have recently very clearly demonstrated that partitioning is a very useful technique for adapting standard matrix computation software to achieve high performance on a variety of parallel and vector machines [5].

Block algorithms, on the other hand, deal with matrices as arrays of smaller matrices. The scalar operations (add, multiply, divide) of a point algorithm become their matrix analogs (matrix-add, matrix-multiply, matrix-inverse). Thus, for example, a block matrix algorithm requires 6 nested loops!

1. Solving Linear Systems

Consider the system

$$Ay = b. \tag{1}$$

where $A \in \mathbf{R}^{n \times n}$. Let A, y, and b have the block structure

$$A = \begin{bmatrix} A_{11} & A_{12} & \cdots & A_{1k} \\ A_{21} & A_{22} & \cdots & A_{2k} \\ \vdots & \vdots & & \vdots \\ A_{k1} & A_{k2} & \cdots & A_{kk} \end{bmatrix}, \quad y = \begin{bmatrix} y_1 \\ y_2 \\ \vdots \\ y_k \end{bmatrix}, \quad b = \begin{bmatrix} b_1 \\ b_2 \\ \vdots \\ b_k \end{bmatrix}. \tag{2}$$

where $A_{ij} \in \mathbf{R}^{b \times b}$, and $kb = n$. If such a factorization of n is not convenient, choose $m = kb > n$ and extend (1) to

$$\begin{bmatrix} A & 0 \\ 0 & I_{m-n} \end{bmatrix} \begin{bmatrix} y \\ 0 \end{bmatrix} = \begin{bmatrix} b \\ 0 \end{bmatrix}.$$

1.1 Block LU Factorization

Compute a factorization $A = LU$ where

$$L = \begin{bmatrix} I & & & \\ L_{21} & I & & \\ \vdots & & \ddots & \\ L_{k,1} & \cdots & L_{k,k-1} & I \end{bmatrix}, \quad U = \begin{bmatrix} U_{11} & U_{12} & \cdots & U_{1,k} \\ & U_{22} & & \vdots \\ & & \ddots & U_{k-1,k} \\ & & & U_{k,k} \end{bmatrix}$$

Here is the algorithm:

Algorithm BLU: Overwrite A with its block LU factorization.

```
for p ← 1 to k
    A_pp ← A_pp^{-1}
    for i ← p + 1 to k
        A_ip ← A_ip A_pp
        for j ← p + 1 to k
            A_ij ← A_ij − A_ip A_pj.
```

This stores

$$\begin{bmatrix} U_{11}^{-1} & U_{12} & \cdots & U_{1,k} \\ L_{21} & U_{22}^{-1} & & \vdots \\ \vdots & & \ddots & U_{k-1,k} \\ L_{k,1} & \cdots & L_{k,k-1} & U_{k,k}^{-1} \end{bmatrix}.$$

in A. To solve eq. (1), we use block forward-solve:

Algorithm BFS: [$L^{-1}b$ overwrites b]

```
for p ← 1 to k − 1
    for i ← p + 1 to k
        b_i ← b_i − A_ip b_p
```

followed by block back-solve:

Algorithm BBS: $U^{-1}b$ overwrites b in

```
for p ← k to 1 step −1
    b_p ← A_pp b_p
    for i ← 1 to p − 1
        b_i ← b_i − A_ip b_p
```

Unfortunately, this algorithm may fail even though A is nonsingular. It may happen that a pivot block A_{pp} is singular or nearly so. If A is diagonally dominant, however, it does not fail. In general, if every leading principle block submatrix of A is nonsingular, it does not fail.

Note that k inversions of $b \times b$ blocks are required, and that $\frac{1}{3}k^3 + O(k^2)$ matrix products are used. Thus, the ratio of operations done in the form of matrix multiply to those done as as matrix inverse is $\frac{1}{3}k^2$; hence a machine whose general purpose speed is at least $\frac{3}{k^2}$ times as fast as its matrix multiply speed will be balanced for this algorithm.

To best allow fast access to the data on machines with interleaved memory, it is advantageous to store A in blocked form. In blocked storage, a block of the matrix occupies b^2 consecutive storage location.

1.2 Inverses of Pivot Blocks

Since they are required by Algorithm BLU, we must ask how to provide inverses of $b \times b$ blocks. There are two reasonable methods.

1. Use a direct method.
2. Use an iterative method due to Schulz (right, no "t")[14] that computes the inverse using matrix multiplications.

Direct computation of the inverse of a $b \times b$ matrix requires $2b^3$ operations in general, and b^3 if the matrix is symmetric positive definite.

1.2.1 The Schulz Method

The method is

$$X_0 \leftarrow \alpha_0 B^T$$

$$X_{i+1} \leftarrow \alpha_{i+1}(2I - X_i B)X_i$$

which costs $4b^3$ operations per step. The sequence $\{X_i\}$ converges to B^+, the pseudo-inverse of B. The number of steps required is about $\log_2 \kappa(B)$ where $\kappa(B)$ is the spectral condition number of B (the ratio of largest to smallest singular values). Since the number of iterations may be considerably greater than 1, this method is very expensive. But in some (admittedly unusual) circumstances this may be acceptable because matrix multiply can be done so fast.

Schreiber [12] gives details on computing the α_i. It suffices for convergence to take $\alpha_i = 1$ for $i > 0$ and

$$\alpha_0 < \frac{2}{\sigma_1^2}$$

where σ_1 is the largest singular value of B. To estimate σ_1^2, we may take $\|B\|_F^2$, since

$$\sigma_1^2 \leq \sigma_1^2 + \cdots + \sigma_b^2 = \|B\|_F^2 = \text{ trace } (B^T B) \leq b\sigma_1^2.$$

Better choices of α_i, which save about half the iterations, are described in the paper [12].

1.3 An Alternative Factorization

We may also compute a partitioned LU factorization

$$A = LU$$

where

$$L = \begin{bmatrix} L_{11} & & 0 \\ \vdots & \ddots & \\ L_{k1} & \cdots & L_{kk} \end{bmatrix}; \quad U = \begin{bmatrix} U_{11} & \cdots & U_{1k} \\ & \ddots & \vdots \\ 0 & & U_{kk} \end{bmatrix}$$

and where the L_{ii} are unit-lower triangular and the U_{ii} are upper triangular. This is especially advantageous if solving a triangular system can be done as fast as matrix multiply.

The algorithm is:

for $p \leftarrow 1$ **to** k
  factor $A_{pp} = L_{pp}U_{pp}$; store the factors in A_{pp}
  for $i \leftarrow p+1$ **to** k
    $A_{ip} \leftarrow A_{ip}U_{pp}^{-1}$
    $A_{pi} \leftarrow L_{pp}^{-1}A_{pi}$
  for $i \leftarrow p+1$ **to** k
    for $j \leftarrow p+1$ **to** k
      $A_{ij} \leftarrow A_{ij} - A_{ip}A_{pj}$

This method is attractive since a factorization costs less than an inversion: $\frac{2}{3}b^3$ operations ($\frac{1}{3}b^3$ if A is symmetric positive definite.)

If solving triangular systems is not especially easy we would store L_{pp}^{-1} and U_{pp}^{-1} in A_{pp}. Since L_{pp}^{-1} is unit-lower triangular, these factors fit in the storage for A_{pp}.

For the factorization above, the ratio of matrix multiply and triangular solve operations to factoring is k^2, which is 3 times better than the ratio achieved by the block LU algorithm.

1.3.1 Inverting the Triangular Factors

We can use a version of the Schulz method due to Pan and Reif [10] to invert triangular matrices.

Let the $b \times b$ matrix $L = [\ell_{ij}]$ be lower triangular. Choose $X_0 = \text{ diag } (\ell_{11}^{-1}, \ell_{22}^{-1}, \ldots, \ell_{bb}^{-1})$. Define X_i by the method:

$$X_{i+1} = (2I - X_i L)X_i \qquad i = 0, 1, \ldots \tag{3}$$

and define the residual matrix R_i by

$$R_i = I - X_i L.$$

It is straightforward to show that

$$X_{i+1} = (I + R_i)X_i \tag{4}$$
$$R_{i+1} = R_i^2. \tag{5}$$

Moreover, R_0 is *strictly* lower triangular. Therefore, $R_1 = R_0^2$ is zero above the second subdiagonal, $R_2 = R_1^2$ is zero above the fourth subdiagonal, and, in general, R_k is zero above the 2^kth subdiagonal. Thus, if $2^k \geq b$, R_k is zero and $X_k = L^{-1}$. For example, if $b = 32$, $R_5 = 0$ and $X_5 = L^{-1}$. Moreover, as R_i becomes more and more sparse, we can exploit its structure to reduce the cost of the method using formulas (4) and (5).

1.4 Block Algorithms for Symmetric Positive Definite Matrices

Let

$$A = \begin{bmatrix} A_{11} & A_{21}^T & \cdots & A_{k1}^T \\ A_{21} & \ddots & & \vdots \\ \vdots & & \ddots & \\ A_{k1} & \cdots & & A_{kk} \end{bmatrix}$$

be an $n \times n$ symmetric positive definite (SPD) matrix. All of the leading principle submatrices of A are also SPD and are at least as well conditioned as is A itself, hence the block triangular factorizations discussed above exist and can be stably computed.

We shall assume that A is stored so that only the blocks A_{ij} $i \geq j$ are available.

If A is SPD then it has a block LDL^T factorization, where

$$L = \begin{bmatrix} I & & & \\ L_{21} & \ddots & & \\ \vdots & & \ddots & \\ L_{k1} & \cdots & & I \end{bmatrix}, \qquad D = \text{ diag } (D_1, \ldots, D_k)$$

We shall overwrite A_{ii} with D_i^{-1} and A_{ij} with L_{ij}, $i > j$.

Note that to then solve (1), we first solve $(LD)y = b$; since

$$LD = \begin{bmatrix} D_1 & & & 0 \\ L_{21}D_1 & \ddots & & \\ \vdots & & \ddots & \\ L_{k1}D_1 & \cdots & & D_k \end{bmatrix},$$

one may solve for y with the following algorithm:

$$\begin{aligned} &\textbf{for } p \leftarrow 1 \textbf{ to } k \\ &\quad y_p \leftarrow D_p^{-1} b_p \\ &\quad \textbf{for } i \leftarrow p+1 \textbf{ to } k \\ &\qquad b_i \leftarrow b_i - L_{ip} D_p y_p \\ &\qquad\quad = b_i - L_{ip} b_p . \end{aligned}$$

Here is the factorization code:

$$\begin{aligned} &\textbf{for } p \leftarrow 1 \textbf{ to } k \\ &\quad A_{pp} \leftarrow A_{pp}^{-1} \\ &\quad \textbf{for } j \leftarrow p+1 \textbf{ to } k \\ &\qquad T \leftarrow A_{pp} A_{jp}^T \qquad (= D_p^{-1} A_{pj}) \\ &\qquad \textbf{for } i \leftarrow j \textbf{ to } k \\ &\qquad\quad A_{ij} \leftarrow A_{ij} - A_{ip} T . \end{aligned}$$

This stores, in A, the blocks:

$$\begin{bmatrix} D_1^{-1} & & & 0 \\ L_{21}D_1 & \ddots & & \\ \vdots & & \ddots & \\ L_{k1}D_1 & \cdots & & D_k^{-1} \end{bmatrix}.$$

2. Least Squares Problems

The problem is $\min_x \|Ax - b\|$. The norm $\| \ \|$ is the 2-norm: $\|y\|^2 = y^T y$. A is $m \times n$, $m \geq n$. We assume that rank $(A) = n$.

2.1 Normal Equations

We may form and solve the *normal equations*: $A^T Ax = A^T b$.

Forming $A^T[A, b]$ is a matrix multiply. (A and b can be stored in an $m \times (n+1)$ array and multiplied by A^T. Then the method for SPD systems (Section 1.4) is used. This will succeed provided A is well-enough conditioned.

2.2 QR Factorization

We seek a factorization

$$Q^T[A, b] = [\tilde{R}, \tilde{b}]$$

where Q is orthogonal and $\tilde{R}$ upper triangular. Partition

$$\tilde{R} = \begin{bmatrix} R \\ 0 \end{bmatrix}, \qquad \tilde{b} = \begin{bmatrix} c \\ d \end{bmatrix}.$$

We then solve $Rx = \tilde{c}$.

2.2.1 Partitioned Givens QR

Define

$$\bar{A} = \begin{bmatrix} A_{11} & \cdots & A_{1s} \\ \vdots & & \vdots \\ A_{r1} & & A_{rs} \\ A_{r+1,1} & \cdots & A_{r+1,s} \end{bmatrix} \qquad m = rb, \quad n = sb$$

where $A_{r+1,j} \equiv 0$. Compute Q and R as follows.

Algorithm PGQR:

for $j \leftarrow 1$ **to** s **begin**
set $R_{r+1,j} \equiv A_{r+1,j}$
for $i \leftarrow r$ **to** j **begin**

$$\begin{bmatrix} R_{ij} \\ 0 \end{bmatrix} \leftarrow Q_{ij} \begin{bmatrix} A_{ij} \\ R_{i+1,j} \end{bmatrix}$$

(where R_{ij} and $R_{i+1,j}$ are upper triangular and Q_{ij} is a $2b \times 2b$ orthogonal product of plane rotations.) Next set

$$\begin{bmatrix} A_{i,j+1} & \cdots & A_{i,s} \\ A_{i+1,j+1} & \cdots & A_{i+1,s} \end{bmatrix} \leftarrow Q_{ij} \begin{bmatrix} A_{i,j+1} & \cdots & A_{i,s} \\ A_{i+1,j+1} & \cdots & A_{i+1,s} \end{bmatrix} \tag{6}$$

end; **end**;

Note that the update (6) is a matrix multiply. Finally, set

$$R = \begin{bmatrix} R_{11} & A_{12} & \cdots & A_{1s} \\ & R_{22} & \cdots & A_{2s} \\ & & \ddots & \vdots \\ & & & R_{ss} \end{bmatrix}.$$

Note that $\text{diag}(L) = (2, \ldots, 2)$. QED

Note that in both of the partitioned QR algorithms, $O(\frac{s-1}{s})$ of the work is matrix multiply, where s is the number of block columns of A.

Bischof and Van Loan [1] have developed a variant of this method that represents the product of several Householder transformations in the form $I - WY$ rather than $I - VLV^T$ as we have done above.

2.2.3 A Block QR Algorithm

We seek a factorization

$$QA = \bar{R} = \begin{bmatrix} R \\ 0 \end{bmatrix}$$

where

$$R = \begin{bmatrix} R_{11} & R_{12} & \cdots & R_{1s} \\ & & \ddots & \vdots \\ 0 & & & R_{ss} \end{bmatrix}$$

and where each $R_{ij} \in \mathbf{R}^{b \times b}$, *i.e.*, R is *block* upper triangular. The other factor $Q = Q_s \cdots Q_1$ where, for example, Q_1 reduces the first block column of A to $b \times b$ form:

$$Q_1 \begin{bmatrix} A_{11} \\ \vdots \\ \vdots \\ A_{r1} \end{bmatrix} = \begin{bmatrix} R_{11} \\ 0 \\ \vdots \\ 0 \end{bmatrix}$$

and Q_1 is a *block reflector*:

$$Q_1 = I - 2V_1V_1^T, \quad V_1 \in \mathbf{R}^{m \times b}, \quad V_1^TV_1 = I_b.$$

In general, if

$$(Q_{j-1} \cdots Q_1)A = \left[\begin{array}{ccc|ccc} R_{11} & \cdots & R_{1,j-1} & R_{1j} & \cdots & R_{1s} \\ & \ddots & & & & \\ & & R_{j-1,j-1} & R_{j-1,j} & \cdots & R_{j-1,s} \\ \hline & & & A_{j,j} & \cdots & A_{j,s} \\ & 0 & & \vdots & & \vdots \\ & & & A_{r,j} & \cdots & A_{r,s} \end{array}\right] \equiv A^{(j-1)}$$

then Q_j is chosen so that

$$Q_j \begin{bmatrix} A_{j,j} \\ \vdots \\ \vdots \\ A_{r,j} \end{bmatrix} = \begin{bmatrix} R_{jj} \\ 0 \\ \vdots \\ 0 \end{bmatrix}$$

The block reflectors are symmetric, orthogonal matrices that are known, in the special case where $b = 1$, as *elementary reflectors* [6].

Parlett and Schreiber [11] describe the construction of the block reflector Q. The method starts by computing P and T such that

$$\begin{bmatrix} A_{jj} \\ \vdots \\ A_{rj} \end{bmatrix} = \begin{bmatrix} P_1 \\ P_2 \end{bmatrix} T = PT$$

where $P^TP = I_r$ and $P_1 \in \mathbf{R}^{r\times r}$, $r \leq b$, and $T \in \mathbf{R}^{r\times b}$. A point-$QR$ factorization may be used to compute P and T.

The method continues by computing

$$P_1 = UM \qquad \text{polar decomposition}$$

where U is orthogonal and M is SPD;

$$(I + M) = R^TR \qquad \text{Cholesky decomposition}$$

where R is upper triangular; and finally

$$V_j = \frac{1}{\sqrt{2}}\begin{bmatrix} UR \\ -P_2R^{-1} \end{bmatrix}.$$

We shall discuss next a method for the polar decomposition that is rich in matrix multiply.

Another method for computing a block-QR decompostion, based on *nonsymmetric* orthogonal matrices of the form $I - WY^T$, with $W, Y \in \mathbf{R}^{m\times b}$ was given by Brønlund and Johnsen [4].

2.3 Polar Decomposition

Because of its use in the block-QR algorithm of the previous section, we consider next algorithms for the polar decomposition of a matrix. The polar decomposition has a number of uses in addition to the one given above [8].

Given an invertible matrix B, the factorization $B = UM$ where U is orthogonal and M is symmetric non-negative definite is the polar decomposition of B. [If b is a complex number not equal to zero, then $b = (b/|b|)|b|$ is its polar decomposition.] Clearly, M is the unique positive definite square root of B^TB and hence $U = BM^{-1}$ is unique too.

Higham [7] has given an algorithm of the form

$$X_0 = f(B),$$
$$X_{i+1} = g(X_i, X_i^{-1}).$$

The sequence $\{X_i\}$ converges (globally, and ultimately quadratically) to the orthogonal polar factor U.

Since the X_i converge quadratically, we have

$$\|X_{i+1} - X_i\| = 0(2^{-2i}).$$

It follows that $\|X_{i+1}^{-1} - X_i^{-1}\| = 0(2^{-2i})$. And, since $X_\infty = U$ is well-conditioned, no large constant is hidden by the $O(\cdot)$.

Almost all of the work in Higham's method is in computing X^{-1} at every iteration. We propose to use Schulz's method with X_{i-1}^{-1} as starting guess. Experiments show that, in a typical case, there are five iterations required for Higham's method to converge, and the number of Schulz iterations is as follows:

iteration	Schulz iterations
1	6
2	5
3	3
4	2
5	1
Total	17

Thus, about 34 matrix products are required to compute the polar decomposition in this way.

3. Eigenvalue Problems for Symmetric Matrices

For every real $n \times n$ symmetric matrix A, there exists a real orthogonal matrix V and real diagonal matrix Λ such that

$$A = V \Lambda V^T.$$

This is the eigendecomposition of A.

We consider a parallel block Jacobi method. It is easy to comprehend the algorithm by observing how it works in a simple case. Suppose that A is a block 4×4 matrix

$$A = \begin{bmatrix} A_{11} & A_{12} & A_{13} & A_{14} \\ A_{12}^T & A_{22} & A_{23} & A_{24} \\ A_{13}^T & A_{23}^T & A_{33} & A_{34} \\ A_{14}^T & A_{24}^T & A_{34}^T & A_{44} \end{bmatrix}$$

where each A_{ij} is $b \times b$.

Algorithm BJ (Block Jacobi):

repeat {

STEP 1:

$$\begin{pmatrix} A_{11} & A_{12} \\ A_{12}^T & A_{22} \end{pmatrix} \leftarrow P_{12} \begin{pmatrix} A_{11} & A_{12} \\ A_{12}^T & A_{22} \end{pmatrix} P_{12}^T$$

where P_{12} is a product of plane rotations representing *one sweep* of a cyclic Jacobi method applied to a $2b \times 2b$ matrix; *i.e.*, P_{12} is a product of $b(2b-1)$ Jacobi rotations.

Also

$$\begin{pmatrix} A_{33} & A_{34} \\ A_{34}^T & A_{44} \end{pmatrix} \leftarrow P_{34} \begin{pmatrix} A_{33} & A_{34} \\ A_{34}^T & A_{44} \end{pmatrix} P_{34}^T$$

where P_{34} is generated similarly.

STEP 2:

$$\begin{pmatrix} A_{13} & A_{14} \\ A_{23} & A_{24} \end{pmatrix} \leftarrow P_{12} \begin{pmatrix} A_{13} & A_{14} \\ A_{23} & A_{24} \end{pmatrix} P_{34}^T.$$

This is matrix multiply.

STEP 3R: Permute block rows of A as follows:

$$A \leftarrow \begin{bmatrix} A_{11} & A_{12} & A_{13} & A_{14} \\ A_{41} & A_{42} & A_{43} & A_{44} \\ A_{21} & A_{22} & A_{23} & A_{24} \\ A_{31} & A_{32} & A_{33} & A_{34} \end{bmatrix}.$$

STEP 3C: Permute block columns in the same way:

$$A \leftarrow \begin{bmatrix} A_{11} & A_{14} & A_{12} & A_{13} \\ A_{41} & A_{44} & A_{42} & A_{43} \\ A_{21} & A_{24} & A_{22} & A_{23} \\ A_{31} & A_{34} & A_{32} & A_{33} \end{bmatrix}$$

} until convergence.

Note: In general the permutation scheme is:

$1 \quad 2 \qquad 3 \quad 4 \qquad 5 \quad 6 \qquad \cdots \qquad k-1 \quad k$

(We assume the block dimension is even.)

A sweep is one pass over all the off-diagonal blocks. Since, for block $k \times k$ matrices there are $k(k = 1)/2$ such blocks, and $k/2$ are zeroed at each iteration of the loop, $k - 1$ iterations is one sweep.

Variants of this method that are somewhat more efficient are discussed by Schreiber [13]. Experiments done by Brent and Luk [3] indicate that 6–10 sweeps suffice for convergence.

A block version of Eberlein's method for the nonsymmetric case was given by Braun and Johnsen [2].

4. Further Work

Here are some interesting open questions in this area.

1. Luk and Park have recently proven that a number of parallel cyclic Jacobi methods are convergent [9]. Can these proofs be extended to the block Jacobi methods of Section 3 above?
2. Is there any merit to *classical* block Jacobi methods, in which large off-diagonal blocks are used, as proposed by Scott, et. al. [15]
3. Block reflectors can be used, in a obvious way, to block tridiagonalize a symmetic matrix, or to reduce a nonsymmetric matrix to block upper-Hessenberg form. Can these reductions be used as the first step in efficient algorithms for the eigenvalue problem?
4. Are there good block methods for symmetric indefinite linear systems? For Vandermonde (or block Vandermonde) systems?

Acknowledgments

I thank Dr. Ralph Schmidt for suggesting several of these problems and for our many discussions on this subject. This work was partially supported by the Office of Naval Research under contract NOOO14-86-K-0610 and by Saxpy Computer Corporation.

References

[1] Christian Bischof and Charles Van Loan. "The WY representation for products of Householder matrices." *Report TR-85-681, Cornell University Computer Science Department*, 1985.

[2] K.A. Braun and Th. Lunde Johnsen. "Hypermatrix generalization of the Jacobi and Eberlein method for computing eigenvalues and eigenvectors of hermitian and non-hermitian matrices." *Computer Methods in Applied Mechanics and Engineering* 4 (1974) 1-18.

[3] R.P. Brent and F.T. Luk. "The solution of singular value and symmetric eigenvalue problems on multiprocessor arrays." *SIAM Journal on Scientific and Statistical Computing* 6 (1985) 69-84.

[4] O.E. Brønlund and Th. Lunde Johnsen. "QR-factorization of partitioned matrices." *Computer Methods in Applied Mechanics and Engineering* 3 (1974) 153-172.

[5] J.J. Dongarra and D.C. Sorensen. "Linear algebra on high-performance computers." In E. U. Schendel, editor, *Proceedings Parallel Computing 85*, 3-32. North Holland, 1986.

[6] Gene H. Golub and Charles F. Van Loan. "Matrix Computations." Johns Hopkins, 1983.

[7] N. J. Higham. "Computing the polar decomposition — with applications." *SIAM Journal on Scientific and Statistical Computing* 7 (1986) 1160-1174.

[8] N.J. Higham. "Computing a nearest symmetric positive semi-definite matrix." *Numerical Analysis Report No. 126, University of Manchester Department of Mathematics*, Nov. 1986.

[9] Franklin T. Luk and Haesun Park. "A proof of convergence for two parallel Jacobi SVD algorithms." *Report EE-CEG-86-12, Cornell University School of Electrical Engineering*, Nov. 1986.

[10] V. Pan and J. Reif. "Efficient parallel solution of linear systems." In *Proceedings of the Seventeenth Annual ACM Symposium on Theory of Computing*, (1985) 143-152.

[11] B. Parlett and R. Schreiber. "Block reflectors: Theory and computation." *SIAM Journal on Numerical Analysis*, to appear.

[12] R. Schreiber. "Computing generalized inverses and eigenvalues of symmetric matrices by systolic array." in R. Glowinski and J. L. Lions, editors, *Computer Methods in Applied Science and Engineering*, 285-295. North-Holland, 1984.

[13] R. Schreiber. "Solving eigenvalue and singular value problems on an undersized systolic array." *SIAM Journal on Scientific and Statistical Computing* 7 (1986) 441-451.

[14] G. Schulz. "Iterative Berechnung der reziproken Matrix." *Z. Angew. Math. Mech.* 13 (1933) 57-59.

[15] David S. Scott, Michael T. Heath and Robert C. Ward. "Parallel block Jacobi eigenvalue algorithms using systolic arrays." *Linear Algebra and Its Applications* 77 (1986) 345-355.

The LCAP System: An Example of a Multicomputer*

Martin H. Schultz

1. Introduction

Suppose we have a supercomputer which is too slow. What can we do to get more speed? It is now very fashionable to think of "going parallel." In this paper we describe one approach called the LCAP system developed by IBM/Kingston and Scientific Computing Associates (SCA). The goals of the LCAP project were to:

1. develop a prototype parallel computer system using standard off-the-shelf supercomputers as nodes so that experimental research in parallel algorithm and programming languages could be initiated as fast as possible;
2. make the system easy to use for noncomputer scientists, which meant using a parallel programming environment consisting of Fortran and simple parallel programming extensions;
3. keep the system simple, relatively inexpensive, and cost effective which meant in practice using FPS-X64 machines as the nodes; and
4. allow incremental extention of the system as more computer power was needed and financial resources were available.

Point (4) is increasingly important in today's real world of insatiable demands for cpu cycles and limited financial resources.

Designing a successful parallel multiprocessor demands that we pay attention to (1) architecture, eg., we need to know what we can build at reasonable cost, (2) parallel algorithms and their analyses, i.e., we need to understand how parallel algorithms of interest perform on different architectures, and (3) programming because we need to implement our parallel algorithms on the parallel architecture. Not all parallel algorithms or parallel architectures may be equally easy to program. Furthermore, it may be possible to modify either the algorithms or architecture to make the programming and debugging significantly easier. Clearly understanding and optimizing the interaction of parallel architectures, algorithms, and programming languages is likely to yield a bigger payoff than optimizing each of these facets of parallel computation in isolation from the others.

2. Architectures

When we consider architectures for scientific computing there are a number of interrelated issues: (1) What is the granularity of the nodes in the parallel system? The range is enormous from bit serial, eg. in the Connection Machine, to vector processors, eg. in the Cray-XMP or ETA10, to tightly coupled vector processors, eg. the Alliant FX-8s in the Ceder system. (2) What is the interconnect among the nodes? The possibilities include, a bus, a ring, a hypercube, a switch. (3) Where is the memory located? Is it all private? Is it essentially all public (shared)? Is it some combination of private and public?

A surprisingly large number of the currently available commercial and even experimental systems are based on a bus or switch or a ring architecture. These systems vary in two very important regards: (1) Do they have shared memory? and (2) If they do have shared memory, do they have significant private memories? By significant we mean that the private memories are large enough to store the data sets required for nontrivial problems in a distributed fashion. We can use the following table to classify some systems:

*This Research was supported in part by ONR Grant N00014-86-K-0310 and IBM/Kingston

	Shared Memory		No Shared Memory
	Small Private Memory	Large Private Memory	Large Private Memory
Ring	Warp Saxpy	LCAP	CYBERPLUS
Bus/ Switch Hypercube	Alliant Encore Sequent Cray	ETA10 LCAP Cedar Flexible	IPSC T-SERIES NCUBE

Table 1:

As we will see, algorithm analysis tells us that if we have enough private memory to store the problem data in a distributed sense, we can often get by with an order of magnitude less data movement than if we have to store the data in public memory. We will see this phenomena in action in our second concrete example of **Section 4**. The reason is that if we assign the location of the data intelligently to take advantage of locality, i.e. the data a processor needs is either in its own memory or the memory of its neighboring nodes, then only a small fraction of the data need be moved among the nodes. This is a very important observation which allows one to think about machines which are efficient even for a large number of powerful vector nodes.

Now we may ask ourselves: What is the simplest, fast interconnect we knew how to build at reasonable cost? After some consideration most people would come back with the answer: A bus. In fact, the interconnect in the IBM LCAP system is based on SCA buses and SCA shared bulk memories. The processors in LCAP1 are FPS-164s while the processors in LCAP2 are FPS-264s. Here are a few facts about the SCA shared bulk memory system: Based on 256K bit parts, each SCA shared bulk memory box can hold up to 128 megabytes. Each box can be connected to up to three distinct SCA buses. Each bus has a bandwidth of 44 megabytes/sec and can support up to four memory boxes and four processors; the limiting factor being largely bus length. Finally each FPS processor can be on a number of different buses. These flexible components provide a "toolbox" which allows us to design a variety of multicomputer systems. The LCAP systems each have 10 processors and 4 SCA memory boxes on three SCA buses to form the global shared bulk memory.

In addition, LCAP1 and LCAP2 have a ring structure which is built by using *locally* connected SCA buses and shared bulk memories, [1, 2, 3]. Thus the LCAP architectures naturally supports either shared memory or ring models of parallel computation.

There are a few additional observations we need to make concerning the LCAP architecture. Because each processor is a fully capable computer it not only can support a large private memory but also a powerful I/O system with mechanical disk systems, data displays, and data acquistition subsystems. This automatically provides us with a parallel I/O system which can be very important in practice. The fact that the shared memories are bulk memories means that they are relatively low cost and therefore large amounts of memory can be afforded. Of course, the price we pay in performance over shared random access memory is that we get effective high transfer rates only for long vectors (as distinct from scalars), there is a nontrivial latency or startup cost in performing an I/O event, but the actual rate of transfer is quite high. However, as we will see in our examples this is usually not a problem in this context. Since we can afford to replicate the bulk memories, they can be used in two fashions: (1) as a private virtual solid state disk for each processor and (2) as a mechanism for storing data which must be shared among the processors. In fact, the LCAP system has SCA software to support both shared memory and message passing protocals.

3. Algorithm General Considerations

Parallel algorithm analysis plays an important role in the assessment of parallel systems. Unfortunately traditional numerical analysis which provides algorithm analyses strictly in terms of arithmetic operation counts is not adequate. Data movement and location must be considered. Thus we need to reconsider virtually all of numerical analysis not only to find and formulate parallel algorithms but to carry out a new kind of analysis.

Since the runtime T_{run} of any parallel algorithm can be bounded above and below by expressions involving $T_{I/O} + T_{CPU}$ where $T_{I/O}$ is the time devoted to moving the data and T_{CPU} is the time devoted to computation in the following manner:

$$1/2\left(T_{I/O} + T_{CPU}\right) \leq max\left(T_{I/O}, T_{CPU}\right) \leq T_{run} \leq T_{I/O} + T_{CPU}, \tag{1}$$

we carry out all algorithm analysis in terms of $T_{I/O} + T_{CPU}$. In effect, we are making the implicit worst case assumption that a node can not simultaneously move data and do computation. Inequality (1) shows that this assumption may be pessimistic by at most a factor of 2. In general algorithm analysis for parallel algorithms yields results of the form:

$$T_{run} = f(k)T + C_1 k^{\alpha} \frac{n^q}{b} + C_2 \frac{n^p}{kS} \tag{2}$$

where the first term is the data communication latency ($f(k)$ is monotonically nondecreasing as a function of k), the second term is the time for data transfers ($-1 \leq \alpha \leq 1$, b = rate of data transfer between a single processor and the shared memory), and the third term is the time for computation (S = rate of computation). In multiprocessors which have only private memories and no capability for pipelining mulithop data communication, $f(k)$ is often determined by the need to broadcast data from one processor to all the others, i.e., $f(k)$ is the "diameter" of the multiprocessor. Domain decomposition is a common technique for developing parallel algorithms for partial differential equations. Geometrical considerations of physical space determine that usually $q = p - 1$ which means that domain decomposition algorithms entail an order magnitude less data communication then computation. Furthermore, one dimensional domain decomposition on multiprocessors without shared memory has $\alpha = 0$.

4. Two Specific Algorithms

A large number of parallel algorithms have been developed, analyzed, implemented and benchmarked for the LCAP systems, [1, 2] . By and large these parallel algorithm efforts have been extraordinarily successful and have yielded very high efficiencies or speedups. In this section we present and analyze two simple algorithms of general interest: (1) two dimensional FFT and (2) the five-point finite difference filter. The first problem illustrates the effective use of the global shared memory while the second problem illustrates the use of the ring architecture if the private memories are large enough. If the private memories are not large enough to store the data, then the global shared memory must be used to store all the data and the efficiency of the resulting algorithms may be inadequate. This will dramatically illustrate the possible deficiency of multiprocessor systems without adequate private memories for some problems. We feel that this may be an important observation as many commercially available systems do not have adequate private memories to store all the data required for large scale scientific computations.

As our first concrete example, we consider the parallel computation of two dimensional fast Fourier transforms (FFTs). These transforms are central to many image processing applications and spectral methods for solving partial differential equations.

We assume that the complex-valued double precision data is given on an $N \times N$ grid and that there is enough private memory so that it is initially distributed by rows to all the processors, i.e., if there are k processors, each processor has N/k of the rows. We remark that in many applications, for example, in the use of spectral methods for time dependent partial differential equations, we have to do many two dimensional FFTs so it is expedient to have the data distributed among the processors if at all possible.

The algorithm is essentially given by the following major steps:

1. In parallel, each processor does $\frac{N}{k}$ one dimensional FFTs in the x-variable on the data in its private memory;
2. In parallel, each processor writes its x-transformed data (as single block row of k $\frac{N}{k} \times \frac{N}{k}$ blocks) to the shared memory;
3. In parallel, each processor reads its appropriate x-transformed block column data from the shared memory (requiring k separate reads each of a $\frac{N}{k} \times \frac{N}{k}$ block);
4. In parallel, each processor does $\frac{N}{k}$ one dimensional FFTs in the y-variable on the data in its private memory.

We remark that if another two dimensional FFT or inverse FFT is now required we treat the variables in the reverse order, i.e., y. and then x, to avoid an extra transpose of the data. Furthermore, since it is a low order term, we will ignore the time required to do the transpose of subblocks in each processor (actuually between steps 3 and 4).

The total time required for this algorithm is bounded by the time required for each processor to do $2 \times \frac{N}{k}$ one dimensional FFTs plus the time required for all processors to write and read to and from the shared bulk memory, i.e.,

$$\text{TIME} \leq \text{TIME(FFT)} + \text{TIME(WRITE)} + \text{TIME(READ)}. \tag{3}$$

In addition, we have the following bounds:

$$\text{TIME(FFT)} \leq 2\frac{N}{k}\frac{(5N \log N)}{S}, \tag{4}$$

where S is the rate at which each processor can do 64-bit arithmetic operations and k is number of processors.

$$\text{TIME(WRITE)} \leq \frac{kT}{k^*} + \frac{2N^2}{B}, \tag{5}$$

where T is the read/write latency to the shared memory k^* is the maximum number of processors that read or write simultaneously and B is the total bandwidth to the shared bulk memory. We remark that $B \leq k^* b$, where b is the maximum I/O bandwidth of each processor, and k^* is dependent on the number and structure of the buses and memory boxes in the shared bulk memory. Our bound for TIME(WRITE) assumes the worst possible case that only one processor can write at a time. Finally we have

$$\text{TIME(READ)} \leq \frac{k^2}{k^*} + \frac{2N^2}{B}. \tag{6}$$

Putting these bounds all together we have

$$\text{TIME} \le \left(\frac{k}{k^*} + \frac{k^2}{k^*}\right) T + \frac{4N^2}{B} + 2\frac{N}{k}\frac{(5N \log N)}{S}. \tag{7}$$

This expression shows that for most reasonable systems the run time at first goes down as we increase k but then reaches a point of diminishing returns and starts to increase. Eventually, system is spending all its time moving data around and not doing any useful computation.

For the LCAP1 system at IBM/Kingston, $b = 5.5$ megawords/sec, $B = 16.5$ megawords/sec, $T = 70$ microseconds, $k = 10$, $k^* = 3$, and $S = 11$ megaflops/sec. This implies that for LCAP1,

$$\begin{aligned} TIME &\le (3.3 + 33.3)\left(70 \times 10^{-6}\right) + \frac{4 \times 10^6}{16.5 \times 10^6} + \frac{10 \times 10^6 \times 10}{10 \times 11 \times 10^6} \text{ seconds} \\ &\approx (2.3 + 242 + 910) \times 10^{-3} \text{ seconds} = 1.16 \text{ seconds}. \end{aligned} \tag{8}$$

This yields a computational efficiency equal to

$$\frac{.910}{1.16} \times 100 \approx 78.5\%, \tag{9}$$

which is quite respectable for 10 very fast processors.

Now we assume the processors have private memories which are large enough to hold only a single row and its transform. Now we need to use the shared memory to hold the data and in essence each processor can read or write only a single row or column at a time which can have a very dramatic effect on the latency part of the total time. In fact we have

$$\text{TIME} = \text{TIME(FFT)} + 2(\text{TIME(READ)} + \text{TIME(WRITE)}) \tag{10}$$

where TIME(READ) is now bounded by

$$\text{TIME(READ)} \le \frac{N^2}{k^*}T + \frac{2N^2}{k^*b} = \frac{N^2}{k^*}T + \frac{2N^2}{B} \tag{11}$$

where $k^* \le k$ is maximum number of processors that can read or write simultaneously. The corresponding bound for the writing phase is given by

$$\text{TIME(WRITE)} \le \frac{N}{k^*}T + \frac{2N^2}{B}. \tag{12}$$

Thus,

$$\text{TIME} \le \left(\frac{N}{k^*} + \frac{N^2}{k^*}\right) T + \frac{4N^2}{B} + 2\frac{N}{k}\frac{(5N \log N)}{S}. \tag{13}$$

Using the same parameters as we used before for the LCAP1 system except for the amount of memory on each processor, we get

$$\begin{aligned} \text{TIME} &\approx \left(333 + 333 \times 10^3\right)\left(70 \times 10^{-6}\right) + 1.15 \text{ seconds} \\ &= 333 \times 70 \times 10^{-3} + 1.15 \text{ seconds} = 23.3 + 1.15 \text{ seconds} \approx 24.5 \text{ seconds}. \end{aligned} \tag{14}$$

This yields an efficiency of 4.7% which is a disaster. We can evaluate the performance of hypothetical type LCAP1 systems for both large and small private memories and for $k^* = 1, 3$, and 10 to get the following table:

	$k^* = 1$	$k^* = 3$	$k^* = 10$
Large Private Memories	56%	78.5%	92%
Small Private Memories	1.3%	4.7%	11.4%

Table 2: Efficiency of a Two Dimensional FFT Algorithm on LCAP1 with 10 Processors

Clearly the results of **Table 2** are a disaster for multiprocessor systems with small private memories and shared bulk memories. There are two fixes for such systems:

1. increase the private memories and do the subblock transposes locally (the LCAP1 solution); or
2. replace the shared bulk memory by shared random access memory to eliminate the large latency.

In option (2) it is actually adequate to use a shared bulk memory with an interface which supports full speed reads or writes into the memory with addresses differing by a constant stride. This capability would allow us to do a transpose of the data "on the fly."

While the second fix takes care of the inefficiencies for the two dimensional FFT and as such may be important, it does not help in improving the efficiency of our second problem that we consider now.

Our second concrete example is the problem of applying a five point filter to each point, u_{ij}, of a $N \times N$ array of real numbers. Specifically, we consider the operation

$$u_{ij} \leftarrow 4u_{ij} - (u_{i-j,j} + u_{i+1,j} + u_{i,j+1} + u_{i,j-1}) \tag{15}$$

for all $1 \leq i, j \leq N$ where we take $u_{o,j} = u_{N+1,j} = u_{j,o} = u_{j,N+1} = 0$ for all $1 \leq j \leq N$.

This is a common kernel of many scientific computations. For example it arises naturally in image processing algorithms, in implicit finite difference approximations to time dependent partial differential equations, and in iterative methods for solving sparse linear systems.

We consider one dimensional domain decomposition for this problem, i.e., we divide the array into k vertical (or horizontal) slices or subdomains and have each processor apply the filter to all the points in one of the subdomains. First we consider the case in which the processors have enough memory so that the data can be stored locally in a distributed sense. Thus, we assume that the data for each subdomain is stored in the appropriate processor's private memory. Clearly when a processor computes the filter for its boundary points it may need to have data from as many as $2N$ points from two other processors. This data is shared by having each processor write its boundary data, which is to be shared, to the shared memory and then read its "neighbors" data that it needs.

We have the following bound:

$$\text{TIME} \leq \text{TIME(FILTER)} + \text{TIME(WRITE)} + \text{TIME(READ)}, \tag{16}$$

where

$$
\begin{aligned}
\text{TIME(FILTER)} &\leq \frac{10N^2}{kS} \\
\text{TIME(WRITE)} &\leq \frac{k}{k^*}T + \frac{2Nk}{k^*b} = \frac{kT}{k^*} + \frac{2Nk}{B} \\
\text{TIME(READ)} &\leq \frac{2k}{k^*}T + \frac{2Nk}{k^*b} = \frac{2kT}{k^*} + \frac{2Nk}{B}.
\end{aligned}
\tag{17}
$$

Hence,

$$\text{TIME} \leq \frac{3kT}{k^*} + \frac{4Nk}{B} + \frac{10N^2}{kS}.$$

For our generic example of LCAP1 ($k^* = 3, k = 10$), we get the following bound in milliseconds for N = 1000:

$$
\begin{aligned}
\text{TIME} &\leq \frac{3 \times 10 \times 70 \times 10^{-3}}{3} + \frac{4 \times 1000 \times 10}{16.5 \times 10^3} + \frac{10 \times 10^6}{10 \times 11 \times 10^3} \text{ milliseconds} \\
&= .70 + 2.4 + 90.9 = 94.0 \text{ milliseconds}
\end{aligned}
\tag{18}
$$

This yields an efficiency of about 97% which is excellent.

Second we consider the case of processors whose memories are too small to hold the $N \times N$ array in a distributed sense. In this case the entire array has to be read from shared memory, filtered and then written back to shared memory. The number of startups in this case depends on the amount of private memory on each processor and will by $0(N^2)$ as $N \to \infty$, which is very unfortunate. Since we don't know the amount of private memory in this case, we'll be content to give a lower bound to the time by ignoring the latency. If we do this we get

$$\text{TIME} \geq \frac{2N^2}{B} + \frac{10N^2}{kS}. \tag{19}$$

For our LCAP1 example, we get the lower bound

$$
\begin{aligned}
\text{TIME} &\leq \frac{2 \times 10^6}{16.5 \times 10^3} + 90.0 \text{ milliseconds} \\
&\approx 121.2 + 90.9 \text{ milliseconds} \approx 222 \text{ milliseconds}
\end{aligned}
\tag{20}
$$

which yields an efficiency of at most about 41%. This is not a good result, although it is not a complete disaster. Moreover, the efficiency can not be improved by upgrading the shared bulk memory to a shared random access memory.

As with example 1 we can complete a table giving efficiencies of a variety of LCAP type systems for $k^* = 1, 3$, and 10. The results appear in **Table 3**.

Table 2 and **Table 3** illustrate the effectiveness of LCAP-like multicomputer parallel systems with large private memories and shared bulk memories. Similar results have been obtained by a number of investigators for other problems, [1, 2, 3].

	$k^* = 1$	$k^* = 3$	$k^* = 10$
Large Private Memories	93%	97.5%	99%
Small Private Memories	20%	41%	71%

Table 3: Efficiency of 5-Point Algorithm on LCAP1 with 10 Processors.

References

[1] E. Clementi, J. Detrich, S. Chin, G. Corongiu, D. Folsom, D. Logan, R. Caltabiano, A. Carnevali, J. Helin, M. Russo, A. Gnudi, and P. Palamidese, *Large-Scale Computations on a Scalar, Vector, and Parallel Supercomputer,* Parallel Computing, North Holland, 5 (1987), pp. 13–44.

[2] P. Leca, Programming Loosely Coupled Multi-FPS System with Message Passing Primitives: Experiment in Implementing A.D.I. Method on LCAP1 System, *Proceedings of the Second International Conference on Supercomputing,* 1987.

[3] Scientific Computing Associates Inc., *Shared Bulk Memory Systems,* Manual, Scientific Computing Associates Inc., 1985.

A BLOCK QR FACTORIZATION SCHEME FOR LOOSELY COUPLED SYSTEMS OF ARRAY PROCESSORS

CHARLES VAN LOAN†

1. Introduction. Computing the QR factorization of a matrix $A \in R^{m \times n}$ involves finding an orthogonal matrix $Q \in R^{m \times m}$ and an upper tringular matrix $R \in R^{m \times n}$ such that $A = QR$. This factorization has a prominent role to play in numerical linear algebra especially because of its bearing on the least square problem. A detailed description of the QR factorization and the various ways that it can be computed may be found in Golub and Van Loan (1983).

Parallel methods for computing the QR factorization have received considerable attention recently. For systolic arrays attention has focused on methods that rely on Givens rotations. See Gentleman and Kung (1981) or Heller and Ipsen (1983). Dongarra, Sameh, and Sorenson (1986) have implemented both parallel Givens and parallel Householder procedures on the Denelcor Hep.

In this paper we discuss a block version of the Gentlemen-Kung method that we have implemented on the IBM Kingston LCAP-1. This system consists of ten FPS-164 array processors (APs) that can communicate through several shared bulk memories. An overview of LCAP-1 is offered in Clementi and Logan (1985). The features of LCAP-1 that figure in the current work are depicted in the following diagram:

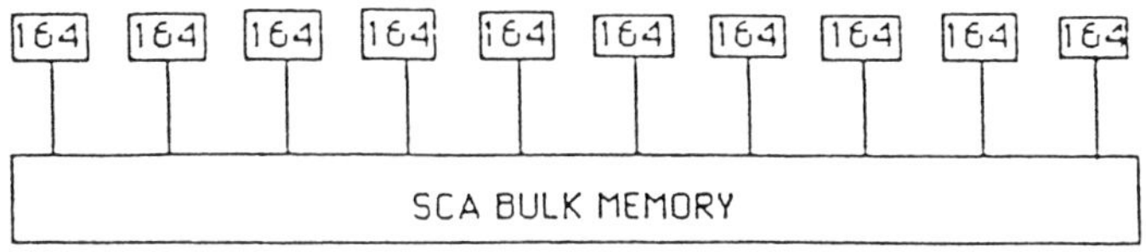

There are actually two levels of parallelism here because the APs are each capable of performing twenty parallel dot products. Indeed, the FPS-164/MAX's at Kingston each come equipped with two "MAX boards". The MAX Board enhancement enables each AP to perform matrix–matrix multiplication at a peak rate of 55 Mflops *if* the matrices involved are sufficiently large. Full exploitation of the FPS–164/ MAX requires having an algorithm that is rich in matrix multiplication. This is why we have chosen to develop a parallel *block* procedure. The blocking of the matrix A is largely a function of the 164/MAX architecture. For example, it turns out to be efficient to have block columns that are a multiple of twenty simply because the LCAP–1 APs can *each* perform twenty parallel dot products. Further details concerning the FPS–164/MAX architecture may be found in Charlesworth and Gustafson (1986).

The matrix A is stored in a 64 Mword bulk memory unit manufactured by Scientific Computing Associates (SCA). Thus, a dense problem of size 16K–by–4K could potentially be solved. The APs have approximately 600 Kwords of usable memory. This is enough to house, for example, a 1000–by–500 submatrix.

Data between the APs and the bulk memory flows at a rate of 44 Mbytes/sec. However, high latency associated with each transferred message demands that data be moved in fairly good–sized chunks in order to be efficient, e.g., 1000 words.

†Department of Computer Science, Cornell University, Ithaca, New York 14853

Additional nuances of the LCAP–1 system as they apply to our QR implementation are detailed later.

This paper is the first of several reports in which we explore the issues associated with parallel matrix computations on the LCAP–1. The parallel block QR factorization scheme that we encoded is derived in §2 and §3. Implementation details are covered in §4 and results in §5. Our current QR code can be improved in several ways as we often opted for the "easy way out" when confronted with an algorithmic dilemma. Despite this we feel that our LCAP–1 experience offers general perspective on large scale distributed matrix computations.

2. Parallel Givens QR. We say that $G \in R^{m\times m}$ is an adjacent Givens rotation in planes i–1 and i if G is the identity with the following 2–by–2 exception:

$$\begin{bmatrix} g_{i-1,i-1} & g_{i-1,i} \\ g_{i,i-1} & g_{ii} \end{bmatrix} = \begin{bmatrix} \cos(\theta) & \sin(\theta) \\ -\sin(\theta) & \cos(\theta) \end{bmatrix} \qquad 2 \le i \le m$$

Notice that G is orthogonal and that premultiplication by G affects just rows i–1 and i. If $x \in R^m$ then it is not hard to determine $(\cos(\theta), \sin(\theta))$ so that $y_i = 0$ if $y = Gx$. These and other Givens rotations issues are discussed in Golub and Van Loan (1983, pp. 43–47).

Adjacent rotations are important because they only combine adjacent rows or columns when applied to a matrix. Moreover, they can be used to compute the QR factorization of a matrix. Assuming $A \in R^{m\times n}$ $(m \ge n)$ we have:

Algorithm 2.1.

```
For j = 1:n
    For i = m : -1 : j+1
        Determine an adjacent Givens rotation G_ij such
            that if y = G_ij^T A(: , j:j) then y_i = 0, i.e., zero a_ij .
        A := G_ij^T A
    end i
end j
```

Upon completion A is overwritten by R and

$$Q = (G_{m,1} \ \cdots \ G_{21}) \ldots (G_{m,n} \cdots G_{n+1,n})$$

Notice that the algorithm computes R column–by–column and that the zeroing within a column proceeds from the bottom up to the subdiagonal. Here is a depiction of the 4–by–3 case:

$$
\begin{matrix} x & x & x \\ x & x & x \\ x & x & x \\ x & x & x \end{matrix} \rightarrow
\begin{matrix} x & x & x \\ x & x & x \\ x & x & x \\ 0 & x & x \end{matrix} \rightarrow
\begin{matrix} x & x & x \\ x & x & x \\ 0 & x & x \\ 0 & x & x \end{matrix} \rightarrow
\begin{matrix} x & x & x \\ 0 & x & x \\ 0 & x & x \\ 0 & x & x \end{matrix} \rightarrow
\begin{matrix} x & x & x \\ 0 & x & x \\ 0 & x & x \\ 0 & 0 & x \end{matrix} \rightarrow
\begin{matrix} x & x & x \\ 0 & x & x \\ 0 & 0 & x \\ 0 & 0 & x \end{matrix} \rightarrow
\begin{matrix} x & x & x \\ 0 & x & x \\ 0 & 0 & x \\ 0 & 0 & 0 \end{matrix}
$$

To indicate the inherent parallelism in this procedure we resort to a slightly larger example and number the a_{ij} in the order that they are zeroed:

x	x	x	x
8	x	x	x
7	15	x	x
6	14	21	x
5	13	20	26
4	12	19	25
3	11	18	24
2	10	17	23
1	9	16	22

$m = 9,\ n = 4$

Recognize that the computation and application of G_{ij} can begin as soon as $G_{i-1,j-1}$ is applied to A. To illustrate this we tabulate the earliest "time step" that $a_{ij}(i > j)$ can be zeroed:

x	x	x	x
8	x	x	x
7	9	x	x
6	8	10	x
5	7	9	11
4	6	8	10
3	5	7	9
2	4	6	8
1	3	5	7

$m = 9,\ n = 4$

With this notation we see in the example that four Givens updates can be performed during the seventh time step: G_{31}, G_{52}, G_{73}, and G_{94}. If we had 4 processors then they could each be assigned one of these tasks.

The parallelism that we have exposed in the above example can be formalized by rearranging the loop indexing in Algorithm 2.1 and noting that $m + n - 2$ timesteps are required.

Algorithm 2.2.

```
For k = 1: m+n-2
    For All j = 1:n
        i = m-k+1+2(j-1)
        if ( i ≤ m & i ≥ j+1)
            Determine G_ij to zero a_ij
            A:= G_ij^T A
        end
    end j
end k
```

The "For All" statement reminds us that all of the updates $A := G_{ij}^T A$ associated with a given time step k are independent and can be performed in parallel.

We point out that G_{ij} can actually be computed "earlier" than we have indicated. For example, in the $(m,n) = (9,4)$ case above, we have assumed that G_{92} is computed as soon as G_{81} has been applied all the way across the matrix. In fact, G_{92} can be computed as soon as G_{81} has been applied to just the second column. For reasons that we given in §4, we have not implemented the "soon as possible" generation of G_{ij}.

Algorithm 2.2 and its natural variants can be mapped nicely onto systolic networks. See Heller and Ipsen (1983).

3. A Parallel Block QR Factorization Method. Some notation is required before a block version of algorithm 2.2 can be specified. Partition $A \in R^{m\times n}$ as follows:

$$A = \begin{bmatrix} A_{11} & \cdots & A_{1q} \\ \vdots & & \vdots \\ A_{p1} & \cdots & A_{pq} \end{bmatrix} \begin{matrix} m_1 \\ \\ m_p \end{matrix} \qquad (3.1)$$
$$\quad\; n_1 \qquad\quad n_q$$

Here, A_{ij} is $m_i - by - n_j$ and we assume that $m_i \geq n_j$ for all i and j. If Q is an orthogonal matrix of dimension $m_{i-1} + m_i$ then we refer to

$$G_I(Q) = \text{ diag } (I_{m_1}, \ldots, I_{m_{i-2}}, Q, I_{m_{i+1}}, \ldots, I_{m_p})$$

as an adjacent "block Given" rotation in block planes $i-1$ and i.

Algorithm 3.1 (Block Givens QR Factorization).

```
For k = 1: p+q-2
    For All j = 1:q
        i = p-k+1+2(j-1)
        if ( i ≤ p & i ≥ j+1 )
            Determine orthogonal Q_ij such that
```

$$Q_{ij}^T \begin{bmatrix} A_{i-1,j} \\ A_{ij} \end{bmatrix} = \begin{bmatrix} R \\ 0 \end{bmatrix} \quad \text{(R upper triangular)}$$

```
            Set G_ij = G_i(Q_ij) and update A := G_ij^T A
        end
    end j
end k
```

This procedure is identical to Algorithm 2.2 except that blocks are zeroed instead of scalars. Upon completion A is overwritten with a block upper triangular matrix R. Unless all the A_{ij} are square, then R will not be upper triangular as a scalar matrix. For example, if the partitioning in (3.1) is defined by $(m_1, m_2) = (3, 3)$ and $(n_1, n_2) = (2, 2)$ then Algorithm 3.1 overwrites A with

$$R = \left[\begin{array}{cc|cc} x & x & x & x \\ 0 & x & x & x \\ 0 & 0 & x & x \\ \hline 0 & 0 & x & x \\ 0 & 0 & 0 & x \\ 0 & 0 & 0 & 0 \end{array}\right]$$

Of course, it is possible to upper triangularize this matrix with further Givens operations, but that is an annoying but necessary follow-up computation.

However, there is a more serious problem associated with rectangular blocks. Consider the example $(m_1, m_2, m_3, m_4) = (2, 3, 3, 8)$, $(n_1, n_2) = (2, 2)$. At the beginning of the second time step A looks like

x	x	x	x
x	x	x	x
x	x	x	x
x	x	x	x
x	x	x	x
x	x	x	x
0	x	x	x
0	0	x	x
0	0	x	x
0	0	x	x
0	0	x	x
0	0	x	x
0	0	x	x
0	0	x	x
0	0	x	x
0	0	x	x

At this stage, Algorithm 3.1 specifies that we only upper triangularize the submatrix $A(3{:}8,1{:}2)$, i.e., the subproblem defined by blocks A_{21} and A_{31}. However, we see from the figure that a significant amount of zeroing in the second block column can take place concurrently. In particular, we could upper triangularize both $A(3{:}8,1{:}2)$ *and* $A(9{:}16,3{:}4)$.

In general, because the "bottom" submatrix A_{ij} in each subproblem is upper triangular, "taller" submatrices can be upper triangularized throughout Algorithm 3.1. In order to rearrange this algorithm so that "maximally tall" subproblems are solved at each stage, we need to drop the fixed row blocking in (3.1). We continue to assume that A has q block columns with widths $n_1 \ldots n_q$. However, instead of imposing a fixed blocking of A's rows we have chosen to determine the "height" of the subproblems through an integer parameter m_0 that satisfies $m_0 \geq n_1$. In our scheme, the subproblems in the first block column involve at most m_0 rows. Maximally tall subproblems are then solved in subsequent block columns at each step. To illustrate, consider the case $m = 100$, $m_0 = 20$, and $(n_1, n_2, n_3, n_4) = (2, 3, 5, 5)$:

Subproblem Row Ranges

Time Step	Column Ranges			
	1:2	3:5	6:10	11:15
1	81:100	-	-	-
2	63:82	83:100	-	-
3	45:64	65:85	86:100	-
4	27:46	47:67	68:90	91:100
5	9:28	29:49	50:72	73:95
6	1:10	11:31	32:54	55:77
7	-	3:13	14:36	37:59
8	-	-	6:18	19:41
9	-	-	-	11:23

In general four integers rowsrt(t, j), rowend(t, j), colsrt(j), and colend(j) are necessary to describe subproblem (t, j), e.g., 29, 49, 3, and 5 for subproblem (5,2). These index arrays and the total number of time steps t_f required can be computed as follows:

Algorithm 3.2. Let m, n, m_0, q and the column partitioning $(n_1, \ldots, n_q)$ be given with $m \geq n$ and $m_0 > n_1$. This algorithm determines t_f and the index arrays colsrt$(1 : q)$, colend$(1 : q)$, rowsrt$(1 : t_f$, 1:q), and rowend$(1 : t_f\ , 1 : q)$.

```
tf = ceiling( max(0,m-m0) / (m0-n1) ) + q
For t = 1 : tf
   if t = 1
      For j = 1:q
         if j = 1
            colsrt(1) = 1
            colend(1) = n1
            rowsrt(t,j) = max( 1 , m-m0+1)
         else
            colsrt(j) = colend(j-1) + 1
            colend(j) = colend(j-1) + nj
            rowsrt(t,j) = m
         end
         rowend(t,j) = m
      end j
   else
      For j = 1:q
         if j = 1
            rowend(t,1) = rowsrt(t-1,1) + n1 - 1
            rowsrt(t,1) = max(1, rowend(t,1)-m0+1 )
         else
            rowend(t,j) = min( rowsrt(t-1,j) + nj - 1 , m )
            rowsrt(t,j) = max( colsrt(j), min( rowend(t,j-1) +1, m))
         end
      end j
   end
end t
```

A couple of comments are in order. In block column 1, the subproblems "climb" at the "rate" $m_0 - n_1$ and so 1+ ceiling$(\max(0, m - m_0)/(m_0 - n_1))$ steps are required to complete the processing of block column 1. Thereafter one block column per time step is completed. This explains the formula for t_f and why we must have $m_0 > n_1$.

In block column j, "serious" computation does not begin so long as rowsrt$(t, j) =$ rowend$(t, j) = m$. After block column j is fully triangularized, rowsrt(t, j) = colsrt(j) and rowend(t, j) = colend(j), conditions that normally signal that there is "nothing to do" in block column j. (An exception occurs when rowsrt$(t, j) =$ colsrt(j) and rowend$(t, j) =$ colend$(j) = m$.)

With subproblems specified by Algorithm 3.2 we can now describe the overall factorization procedure.

Algorithm 3.3 (Maximally Tall Block Givens QR Factorization).

Given m, n, m_0, q, the column partitioning $(n_1, \ldots, n_q)$ with $m \geq n$ and $m_0 > n_1$, the following algorithm overwrites $A \in R^{m \times n}$ with upper triangular $R = Q^T A$ where Q is orthogonal.

```
Compute t_f , rowsrt(1:t_f ,1:q) , rowend(1:t_f ,1:q),
    colsrt(1:q), and colend(1:q) using Algorithm 3.2
For t = 1 : t_f
    For j = 1:q
        i1 = rowsrt(t,j)
        i2 = rowend(t,j)
        j1 = colsrt(j)
        j2 = colend(j)
        if ( i1 = i2 = m  or ( i1 = j1 & i2 = j2 & j2 ≠ m ) )
            "Nothing to do."
        else
            Compute:  A(i1:i2,j1:j2)  = QR .
            Apply: A(i1:i2,j1:n)  :=  Q^T A(i1:i2,j1:n)
        end
    end j
end t
```

4. Implementation. In this section we discuss three issues associated with the implementation of Algorithm 3.3 on the LCAP-1 system: how A is arranged in shared memory, how the subproblems are solved, and how block column tasks are mapped onto processors.

The Storage of A. At time step t, the relevant row and column delimiters for the $j-th$ subproblem are $i_1 = \text{rowsrt}(t,j)$, $i_2 = \text{rowend}(t,j)$, $j_1 = \text{colsrt}(j)$, and $j_2 = \text{colend}(j)$. Here is what the array processor in charge of this subproblem must accomplish:

(1) Read $A(i_1 : i_2, j_1 : j_2)$ from shared memory.

(2) Compute an orthogonal Q such that $Q^T A(i_1 : i_2, j_1 : j_2) = R$ is upper triangular.

(3) Write the updated $A(i_j : i_2, j_1 : j_2)$ back into shared memory.

(4) Read $A(i_1 : i_2, j_1 + 1 : n)$ from shared memory.

(5) Apply Q^T to $A(i_j : j_2, j_1 + 1 : n)$.

(6) Write the updated $A(i_1 : i_2, j_2 + 1 : n)$ back into shared memory.

We assume that $A(i_1 : i_2,, j_1 : j_2)$ can fit into local memory but that because of its size, the processing of $A(i_1 : i_2, j_2 + 1 : n)$ may have to proceed in "chunks". That is, steps 4–5–6 may have to be repeated with a manageable segment of columns from $A(i_j : i_2, j_2 + 1 : n)$ each time. Note that Q stays in the AP during this process. Because one AP is responsible for applying a given Q, there is no need to pass Q on to another AP.

There is an overhead associated with traffic to and from shared memory. Reads and writes to shared memory are accomplished with a "**move**" command and can only involve continuous portions of memory. Using **move** to transfer n floating point words takes

$$T(n) = (100 + 8n/44) \quad \mu\text{sec}$$

Note that the 100 μsec startup degrades the 44mb/sec peak transfer rate. Thus, a vector of length 1000 takes 281 μsec to move for an effective data transfer rate of 28 mb/sec.

From the standpoint of processing the subproblem at hand, it would be ideal if $A(i_1 : i_2, j_1 : n)$ was continuous in shared memory for then a minimum number of **moves** would be required to carry out steps 1,3,4, and 6 above. For example, to read a contiguous 1000-by-500 submatrix from shared memory would require $T(500,000) = .09$ sec ($\equiv$ 44mb/sec). Unfortunately, storing by blocks in Algorithm 3.3 would impose significant buffer requirements and some tedious data manipulation within each AP. The buffer issue is fairly important because the AP's we used have limited local memory ($\approx$ 600 Kwords).

Because we didn't want addditional buffer requirements to limit further the size of "working" memory we chose to store A in column major order. This implies that r **moves** are required to move a submatrix with r columns. Thus, to read a 1000–by–500 submatrix requires $500 \cdot T(1000) = .14$ sec ($\equiv$ 28 mb/sec). This is actually a typical size for a submatrix move in our algorithm. When the overall implementation is considered, we can easily live with a 28 mb/sec data transfer rate.

Subproblem Solution. The basic computation in Algorithm 3.3 consists of computing a QR factorization and then applying the resulting orthogonal matrix to the "rest of A". The normal "Linpack" way to compute a QR factorization of a matrix $C \in R^{m_0 \times n_0}$ is to use Householder matrices. A Householder matrix is an orthogonal transformation of the form

$$P = I - 2vv^T \qquad v \in R^{m_0}, \; \|v\|_2 = 1.$$

In the Linpack QR procedure Householders $P_1, \ldots, P_{n_0}$ are generated so that $P_{n_0} \cdots P_1 C = R$ is upper triangular. Note that $Q = P_1 \cdots P_{n_0}$.

We now consider the computation $Q^T B$ where B is some matrix. If Q is represented as a product of Householders, then the resulting algorithm is "rich" in matrix–vector multiplications. This is fine for many architectures. However, to exploit fully the FPS–164/MAX architecture, we need an update algorithm that is rich in matrix–matrix multiplication. We could accomplish this by explicitly forming the product $Q = P_1 \cdots P_{n_0}$ before applying it to B. But this would be very costly since $m_0 >> n_0$ usually. An unacceptably large $m_0 - by - m_0$ buffer would also be required by this approach.

Instead, we have chosen to use the "WY" representation for products of Householder matrices that is developed in Bischof and Van Loan (1985). In this scheme $m_0 - by - n_0$ matrices W and Y are generated such that

$$Q = P_1 \cdots P_{n_0} = 1 + WY^T$$

The ensuing update $B := Q^T_B = (1 + WY^T)^T B = B + Y(W^T B)$ is then obtained by a pair of matrix–matrix multiplications:

(i) $$Z = W^T B$$

(ii) $$B = B + YZ$$

For (i) we used the "MAX" routine **pdot** that can compute twenty parallel dot products. To initiate the parallel dot product the relevant twenty vectors must be placed in the MAX registers using another MAX routine called **ploadd**. We examine this in some detail so that an appreciate of MAX board computing can be obtained. Assume that W and Y are $m_0 - by - n_0$ and that n_0 (for simplicity) is a multiple of twenty. If B is $m_0 - by - k$ then here is how the matrix $Z = W^T B$ is formed:

```
For j = 1:20:n0
    Load W(1:m0 , j:j+19)_. in to the max registers using
        pload.
    For i = 1:k
        Compute Z(j:j+19,i:i)  = W(1:m0 , j:j+19) TB(1:m0,i:i)
        using pdot.
    end i
end j
```

The times required for each **pload** and **pdot** are approximately

$$\begin{array}{lll} \textbf{ploadd}: & L(m_0) = 23 + 58.2^* m_0 & (\mu\text{sec}) \\ \textbf{pdot}: & D(m_0) = 29.7 + .738^* m_0 & (\mu\text{sec}) \end{array}$$

Thus, $Z = W^T B$ is obtained in $(n_0/20)(L(m_0) + k \cdot D(m_0))\mu$sec. Since $Z =$ requires $2m_0 n_0 k$ flops, a calculation shows that the effective performance in megaflops is approximately given by

$$\text{Mflop}(W^T B) = \frac{55}{1 + 40/m_0 + 79/k + 31/m_0 k}$$

This expression reveals the penalty for short vectors (small m_0) and for low re–use (small k). Here is a table of some representative Mflop($W^T B$) values:

	k = 100	k = 500	k = 1000	k = 5000
m_0 = 100	25	35	37	39
m_0 = 500	29	44	47	50
m_0 = 1000	30	46	49	52
m_0 = 2000	30	47	50	53

Table 4.1

We mention that because the MAX registers can handle vectors up to length 2047, the subproblem height parameter m_0 should be chosen so that rowend(t, j) - rowsrt$(t, j) \leq 2047$ for all t and j.

We now turn our attention to the rank–n_0 update $B \leftarrow B + YZ$ that makes up the second half of the $B \leftarrow (1 + WY^T)^T B$ computation. For this calculation the FPS–164/MAX has a parallel saxpy capability that appears well suited. With two MAX boards it is possible to perform nine saxpys of the form $c_i \leftarrow c_i + s_i y$ in parallel. Note that this is a rank-one update: $C \leftarrow C + ys^T$. Here is how the update of B would proceed using the parallel saxpy routine **pvsma** and the attending load/unload routines **ploadv** and **punldv**. For simplicity, assume that k is a multiple of 9.

```
For j = 1:9:k
    Use ploadv to load B(1:m0,j:j+8)  into the max registers.
    For i = 1:n0
        Use pvsma to perform the update
            B(1:m0,j:j+8)  ← B(1:m0,j:j+8)  + Y(1:m0,i:i)Z(i:i,j:j+8)
    end i
    Use punldv to write the updated B(1:m0,j:j+8)  back to memory.
end j
```

Reasoning as we did to determine Mflop(W^TB), it can be shown that

$$\text{Mflop}(B+YZ) = \frac{24}{1+34/m_0+70/n_0+62/m_0n_0}$$

Note that the re–use factor is now n_0 rather than k. This is unfortunate since in our application we typically have $k > m_0 \gg n_0$. If we look at some typical values of Mflop($B+YZ$), then this is what we find:

	$n_0 = 20$	$n_0 = 40$	$n_0 = 60$	$n_0 = 80$
$m_0 = 100$	4.9	7.7	9.5	10.8
$m_0 = 500$	5.2	8.5	10.7	12.3
$m_0 = 1000$	5.3	8.6	10.9	12.6
$m_0 = 2000$	5.3	8.6	11.0	12.7

Table 4.2

Thus, **pvsma** is ill–suited for the $B \leftarrow B + YZ$ update when compared to the 23–53 Mflop rates sustained by the **pdot** computation of $Z = W^TB$. For this reason we chose to use a new FPS parallel matrix multiply routine called **pmmul** that can perform the update $B \leftarrow B + YZ$ at rates more consistent with the values in Table 4.1

Two final comments about subproblem solution. The first concerns the recording of the orthogonal matrix Q. This matrix is the product of Householder matrices. Of course, these Householders are clustered and applied in WY form during Algorithm 3.3. But we can save all the Householder vectors by overwriting each zeroed subcolumn of A by the corresponding Householder vector. In particular, whenever a subcolumn $v \in R^d$ of A is zeroed by a Householder matrix $(1+2uu^T/u^Tu)$, we store $u(2:d)$ in $v(2:d)$ with the convention $u(1) = 1$. It is then possible to retrieve Q from the final array A so long as the index arrays rowsrt, rowend, colsrt, and colend are available.

Lastly, we mention that the subproblem QR factorizations in Algorithm 3.3 are typically of matrices that have a band structure. Indeed, it is usually the case that A(rowsrt(t,j):rowend(t,j), colsrt(j):colend(j)) has lower bandwidth rowsrt$(t-1,j)$–rowsrt(t,j). This fact is exploited when the QR factorization is computed and the resulting WY factors found.

Load Balancing and Scheduling. Suppose Algorithm 3.3 is to be implemented on array processors $\text{AP}_1,\ldots,\text{AP}_p$. At time step t in Algorithm 3.3 there are q independent tasks to perform.

Task (t,j) involves

$$\text{Factoring:} \quad A(\text{rowsrt}(t,j):\text{rowend}(t,j), \text{colsrt}(j):\text{colend}(j)) = QR$$
$$\text{Computing:} \quad Q^T A(\text{rowsrt}(t,j):\text{rowend}(t,j), \text{colsrt}(j):n))$$

Here, t and j satisfy $1 \le t \le t_f$ and $1 \le j \le q$. If $p = q$ then an immediate load balancing problem arises if each block column has the same width because task (t,j) generally has more matrix to update than task $(t,j+1)$. One way around this difficulty is to make each block column wider than its predecessor. We illustrate this for the case $q = 2$ with block column widths n_1 and n_2. Assuming a subproblem height of m_0 then approximately $2m_0n_1^2 + 2n_1m_0n_2$ flops are required for task $(t,1)$. On the other hand, $2m_0n_2^2$ flops are required for task $(t,2)$ if we again assume a subproblem height of m_0. These two flop counts are approximately equal if $(n_1/n_2) \approx .62$.

For general q it is possible to work out quotients n_j/n_{j+1} for $j = 1:q-1$ so that approximate load balancing results for the column partitioning $n_1, \ldots, n_q$. Of course, in practice it would make more sense to base column partitioning guidelines upon benchmarks rather than upon flop counts. We have not pursued this.

Instead we make the block column widths narrow enough so that the number of independent tasks q is significantly larger than the number p of assigned APs. Approximate load balancing is then achieved by assigning AP$_k$ to block columns $j = k:p:q$. For example, if $p = 3$ and $q = 12$, then AP$_1$ works on block columns 1, 4, 7 and 10, AP$_2$ is assigned to block columns 2,5,8, and 11, while AP$_3$ is applied to block columns 3,6,9, and 12. In a typical time step, each AP will work on 4 subproblems with a greater balance of work than if $q = 3$. This style of distributing tasks has been widely used in parallel matrix factorization work, see George, Heath, and Liu (1985). A fringe benefit of this approach is that we can choose block column widths to be a multiple of twenty. This allows for efficient exploitation of the 164/MAX architecture that permits twenty parallel dot products. In our examples we used uniform block column widths of twenty and thus $q \approx n/20$.

To actually execute algorithm 3.3 in parallel on LCAP–1 we implemented a lock– step synchronization scheme using "barriers". The blocking arrays rowrst, rowend, colsrt, and colend are determined by the host and then downloaded into the p array processors assigned to the computation. The matrix A is also downloaded into the shared memory through the APs. The AP$_k$ then executes the following program:

Algorithm 4.1 (Processor k's Share of Algorithm 3.3).

```
For t = 1:tf
   For j = k:p:q
      Compute  A(rowsrt(t,j):rowend(t,j),colsrt(j):colend(j))  = QR
      Update A(rowsrt(t,j):rowend(t,j),colsrt(j):n)
   end j
   Barrier
end t
```

When the barrier is encountered, execution is suspended until all the other AP programs reach their barrier. After this is accomplished the processing of the next time step begins.

Further details about the LCAP–1 system software required by our implementation may be found in Chin and Lorenzo (1986).

Some Results and Conclusions. In testing our implementation we ran our codes on random matrices $A \in R^{m\times n}$ with the property that $A(1:m, 1:n-1)e = A(1:m, n:n)$ where e is the vector of all ones. The correctness of R was then confirmed by checking the equations $R(1:n-1, 1:n-1)e \quad = R(1:n-1, n:n)$ and $R(n,n) = 0$.

We report on two of the several examples that we solved using the parallel QR code. We do not pretend that our results are conclusive. They merely confirm some natural suspicions and point the way to future research.

The first example indicates that we can get away with our lock step, coarse grained approach if A is large enough and suitably blocked. Here is what we found by using one, two, and three APs to solve an $(m,n) = (5000, 1000)$ problem with $m_0 = 1000, q = 50$, and $n_1 = \cdots = n_{50} = 20$.

Number of Processors	Time (seconds)	Speed-Up	Effective Mflop
1	606	1.00	17
2	310	1.95	33
3	211	2.87	48

Table 5.1

About 25% of the elapsed time is spent on transmitting submatrices to and from the shared memory. To see roughly where this percent comes from consider the update $B \leftarrow (1 + WY^T)^T B$ of a $1000 - by - 500$ submatrix B in shared memory where $W, Y \in R^{1000\times 20}$. If this update is performed at a rate of 30 Mflops then approximately 1.3 seconds must be devoted to computation. To transfer B to or from shared memory requires about .14 seconds. Thus, the fraction of time spent on communication is approximately $.18 \approx .28/1.58$.

We next discuss an example where the load balancing is not quite so nice, resulting in a degradation of performance. In the example $m = 5040, m_0 = 1040, n = 500, q = 13$, and $n_1 = \cdots = n_{12} = 40, n_{13} = 20$. Three APs were used and thus block column tasks are assigned as follows:

$$\mathrm{AP}_1 \leftarrow (1,4,7,10,13) \qquad \mathrm{AP}_2 \leftarrow (2,5,8,11) \qquad \mathrm{AP}_3 \leftarrow (3,6,9,12)$$

Because only five steps are required to process each block column, there are never more than five "active" tasks at any one time step. This makes load balancing a little problematical. The following table indicates the time (in seconds) that each AP spends computing at each timestep.

Time Step	AP_1	AP_2	AP_3
1	3.52	0.00	0.00
2	3.64	3.15	0.00
3	3.64	3.39	2.82
4	3.64	3.42	3.16
5	5.85	3.39	3.16
6	3.10	5.31	4.83
7	2.70	2.82	4.83
8	2.70	2.46	2.58
9	3.88	2.46	2.24
10	2.05	3.42	2.24
11	1.76	1.79	3.01
12	1.76	1.52	1.54
13	1.99	1.52	1.30
14	.62	1.52	1.30
15	.43	1.61	1.30
16	.43	0.00	0.13
17	.43	0.00	0.00

Table 5.2

The time required for the entire computation is 51.2 seconds, the sum of the maximum times in each row of the table. If computation was equally shared at each time step then approximately 38.1 seconds would be required for the complete computation.

The somewhat inefficient use of the APs highlighted by the second example could be rectified in several ways:

1 Choose a smaller m_0. This would have the effect of increasing the number of tasks to be shared at each time step.

2 Vary the block column widths so as to even out the update work.

3 Instead of letting the AP that generates a Q be entirely responsible for its application, share the update.

We have not fully explored these possibilities. Note that the first and third suggestions imply smaller matrix multiplications and thereby reduced 164/MAX performance.

A more promising way to address the load balancing issue would be to incorporate a dynamic scheduling of tasks as is discussed, for example, in George, Heath, and Liu (1985) and Dongarra, Sorenson, and Sameh (1986). One way to do this is to order the tasks (t, j) defined in §4 as follows:

$$(1,1),(1,2),\ldots,(1,q),(2,1),(2,2),\ldots,(2,q),\ldots,(t_f,1),\ldots,(t_f,q)$$

After completing a task each AP would go to this list and "grab" the next available task subject to rules that preserve the integrity of the overall procedure. We will report on this elsewhere.

Acknowledgements. The implementation of the algorithm described in this paper would not have been possible without the expert assistance of Dr. Doug Logan at IBM Kingston and my Ph.D. student Chris Bischof at Cornell.

This work has been supported by ONR contract N00014–83–K–0640, the Mathematical Sciences Institute at Cornell which is sponsored by the Army Research Office, and by the IBM Corporation. Computations were performed at IBM Kingston and on the Production Supercomputer Facility at Cornell which is supported in part by the National Science Foundation and IBM.

REFERENCES

C. Bischof and C. Van Loan, *The WY Representation for products of Householder matrices*, (1987), to appear in Siam J. Scientific and Statistical Computing.

A.E. Charlesworth and J.L. Gustafson, *Introducing replicated VLSI to supercomputing: the FPS-164/ MAX scientific computer*, (1986), IEEE Computer, March, 10–23.

S. Chin and D. Lorenzo, *Parallel Computation on the Loosely coupled array of processors: tools and guidelines*, (1986), Report KGN-25, IBM Kingston, Dept. 48B, Bldg 963, Kingston, NY 12401.

E. Clementi and D. Logan, *Parallel processing with the loosely coupled array processor system*, (1985), Report KGN-43, IBM Kingston, Dept. 48B, Bldg 963, Kingston, NY 12401.

J. Dongarra, A.H. Sameh and D.C. Sorenson, *Implementation of some concurrent algorithms for matrix factorization*, (1986), Parallel Computing, 3, pp. 25–34.

W.M. Gentleman and H.T. Kung, *Matrix triangularization by systolic arrays*, (1982), Proc. SPIE Vol. 298, Real Time Signal Processing IV, 19–26.

A. George, M.T. Heath, and J. Liu, *Parallel Cholesky factorization on a multiprocessor*, (1985), Report ORNL 6124, Math. Sci. Div., Oak Ridge National Laboratory, Oak Ridge, TN.

G.H. Golub and C. Van Loan, *Matrix Computations*, (1983), The Johns Hopkins University Press, Baltimore, Md.

D. Heller and I.C.F. Ipsen, *Systolic networks for orthogonal decompositions*, (1983), Siam J. Scientific and Stat. Computing, 4, 261–269.